BIRKHÄUSER

Klaus Eichmann

The Network Collective

Rise and Fall of a Scientific Paradigm

Birkhäuser
Basel · Boston · Berlin

Prof. Dr med. Dr. h.c. Klaus Eichmann
Max-Planck-Institute of Immunobiology
Stübeweg 51
D-79108 Freiburg

Library of Congress Control Number: 2008930062

Bibliographic information published by Die Deutsche Bibliothek
Die Deutsche Bibliothek lists this publication in the Deutsche Nationalbibliografie;
detailed bibliographic data is available in the Internet at <http://dnb.ddb.de>.

ISBN 978-3-7643-8372-5   Birkhäuser Verlag AG, Basel – Boston – Berlin

The use of registered names, trademarks etc. in this publication, even if not identified as
such, does not imply that they are exempt from the relevant protective laws and regula-
tions or free for general use.
This work is subject to copyright. All rights are reserved, whether the whole or part of
the material is concerned, specifically the rights of translation, reprinting, re-use of illustrations, recitation, broadcasting, reproduction on microfilms or in other ways, and storage in data banks. For any kind of use, permission of the copyright owner must be
obtained.

© 2008 Birkhäuser Verlag AG, P.O. Box 133, CH-4010 Basel, Switzerland
Part of Springer Science+Business Media
Printed on acid-free paper produced from chlorine-free pulp. TCF ∞
Printed in Germany
ISBN: 978-3-7643-8372-5   e-ISBN 978-3-7643-8373-2

9 8 7 6 5 4 3 2 1   www. birkhauser.ch

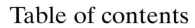

Table of contents

Acknowledgements vii

Chapter 1.   Autobiographical note 1

Part I. Scientific Knowledge, Delusive or Deductive

Chapter 2.   Realism, constructivism, and the naiveté of the
             experimental scientist 7
Chapter 3.   Beyond underdeterminism: Popper, Kuhn, et al. 17
Chapter 4.   The anthropology of science: Ludwik Fleck et al. ... 26
Chapter 5.   The science wars 38
References and further reading, Part I 43

Part II. Origins, Rise, and Fall of the Network Paradigm

Chapter 6.   The immune system, pre-network paradigms 51
Chapter 7.   The necessity for an interactive theory of immunity . 62
Chapter 8.   Proto-ideas of the network theory: antibody
              self-regulation, idiotypy, the brain analogy,
              and cybernetics 72
Chapter 9.   The idiotypic network theory 82
Chapter 10.   The T cell receptor puzzle 95
Chapter 11.   Suppression turned idiotypic 106
Chapter 12.   Network mannerism 116
Chapter 13.   Post-network immunology:
              Idiotypic network continues at the bedside 124
Chapter 14.   Hindsight: Personal interviews 133
              C. Bona, p. 134; P.A. Cazenave, p. 138; A. Coutinho,
              p. 142; R. Germain, p. 149; H. Köhler, p. 155;
              W. E. Paul, p. 158; K. Rajewsky, p. 162; D. Sachs, p. 167;
              E. Sercarz, p. 170; J. Urbain, p. 172; H. Wigzell, p. 175

References and further reading, Part II 180

Part III. Science between Fact and Fiction

Chapter 15. The fictional nature of scientific notions 207
Chapter 16. Fiction turned fact: The case of antibodies 213
Chapter 17. The enticing network: Fiction forever 225
Chapter 18. Logic and laws in life science 234
References and further reading, Part III 241

Appendix

Repeatedly mentioned or quoted individuals 249

Index   269

Acknowledgements

Writing this book would not have been possible without the help of many well-meaning colleagues, friends, and acquaintances from within and without the scientific world. My thanks go collectively to those who have shared with me their views on science in general and life sciences in particular. When I outlined my plans for this book I received much encouragement but also words of caution. In particular, I was warned not to feed arguments to some characters out there who know little about science but only wait for critical texts such as this one to exploit them in their facile attempts to derogate science. While I take these points very seriously, I decided that I have to take that risk. I am convinced that there is no need for science to camouflage the ways in which it really works. On the contrary, these are the ways by which all that valuable knowledge on nature has been acquired – the only ways by which it could have been acquired.

I am particularly grateful to Rose Black, librarian and a wizard in retrieving ancient scientific literature from sources that only she knows. Uta Völkel, physicist, philosopher, and my sister, helped me in trying to understand science theory, and I am deeply grateful for her invaluable advice and suggestions concerning the science-theoretical parts of this text. My most sincere thanks are reserved for those of my contemporary network researchers who agreed to contribute to this book by giving me extended interviews, and letting me use nearly unedited verbatim transcripts of their spontaneous reflections: Constantin Bona, Pierre-André Cazenave, Antonio Coutinho, Ron Germain, Heinz Köhler, Bill Paul, Klaus Rajewsky, David Sachs, Eli Sercarz, Jacques Urbain, Hans Wigzell. I respect the feelings of those who did not respond to my inquiry. Why bother with errors long past? I apologize to those who were not asked to contribute, particularly the renowned network critics whose arguments are well represented in the literature, and explicitly cited in the book. I also apologize to those colleagues whose work I did not mention or quote, or whose work I discussed in ways they do not approve of. This is a personal and biased account of the network paradigm, not a scientific review. Finally, I am grateful to my editor, Beatrice Menz, for her continuing encouragement and constructive critique during the preparation of the manuscript.

June 2008   Klaus Eichmann

Chapter 1
Autobiographical note

There is a traditional tendency to chase and hound the "real" meaning of texts. ...defenders and critics alike continue to argue over "what is actually intended" by its authors. As a welcome relief from this spectacle, literary theory has increasingly disavowed this kind of text criticism. The current trend is to permit texts a life of their own. ... "what the text says", "what really happened", and "what the authors intended" are now very much up to the reader. It is the reader who writes the text.
B. Latour and S. Woolgar (1986)
Laboratory Life, postscript

Now, about three years after having finished my career as an experimental scientist, I am reflecting what I did during this nearly uninterrupted period of 40 years. Did any of the results I obtained achieve the status of lasting scientific fact? Did I contribute to the knowledge in my field, immunology, or did most of it disappear in the vast realms of research gone by, forgotten, superseded by new, more enthralling information?

The first two years of my post-experimental life I have spent investigating the history of an undoubtedly lasting scientific achievement, the invention of the technique for producing hybridomas that secrete monoclonal antibodies, by my late colleague Georges Köhler together with his postdoctoral supervisor Cesar Milstein[1]. Not only are monoclonal antibodies useful for many practical applications in biotechnology and medicine, including diagnoses and therapies of many diseases. In addition, the fact that a monoclonal antibody is produced by a hybrid between a single antibody-producing B cell and a tumor cell confirmed beyond any reasonable doubt one of the most important theories in immunology, the clonal selection theory[2]. Hybridomas have moreover provided the critical experimental material for elucidating the mechanisms that govern the genetic control of antibody synthesis and diversity, a subject that had been highly controversial for decades. Today, everyone with basic laboratory skills and equipment can make hybridomas, and monoclonal antibodies have revolutionized the whole of biology and medicine in an unprecedented and irreversible way.

Did any of the results that I obtained and published in more than 200 papers during my 40 years as an experimentalist even vaguely approach the quality of Köhler's and Milstein's contribution? Clearly not, but in this respect my contributions are not very different from that of the majority of my scientist colleagues. Exceptional scientific achievements are rare and, as I tried to delineate in *Köhler's Invention*, arise as a result of a set of conditions most of which are uncontrollable by the planning intellect of the experimental scientist. An individual scientist is therefore rarely alone responsible for an outstanding achievement, or for the lack of such, for that matter.

What bothers me, however, is that I spent a significant part of my experimental life in a scientific field that is now "after a glory period, ... nearly forgotten"[3], "just a footnote in the history of immunology"[4], a research area of which "it is safe to say that we never learned anything from it"[5]. Originally formulated by Niels Jerne in 1974, the idiotypic network theory initially postulated idiotypic interactions (see chapter 8) within the highly diverse antibody population to be inevitable in the immune system, likely having some consequences for its function as a dynamic network of interconnected elements (see chapter 9)[6,7]. It then developed into the "network paradigm", as I will call it in this essay, i.e. it was stretched and amplified to explain the development and regulation of all elements of the immune system, including that of antibodies, of T and B lymphocytes, and soluble mediators, etc. The idea rapidly fascinated many immunologists who engaged themselves in experiments aimed at proving the network nature of the immune system. Initial experiments yielded exciting results, and recruited ever more experimentalists and theoreticians into idiotypic research, all contributing to what seemed to have become not only a unifying systems theory of adaptive immunity, but reaching far beyond as well. Between 1974 and today around 5,000 journal articles appeared, dealing with the subject directly or indirectly, with a peak production of more than 300 papers per year between 1982 and 1989. Thereafter, some of the experimental evidence collapsed and a number of interpretations became untenable. Simultaneously, new technologies gave rise to new ideas and observations, and the network theory was put aside as useless. Today, less than 10 papers per year deal with network theory which is largely forgotten in mainstream immunological research. "Today the tidal wave of regulation *via* idiotype networks has receded behind an empty beach ... and we have no idea what it was that produced the tidal wave and its ebb"[5].

I was among the first to join in. Between 1974 and 1986 I published 31 original papers on or related to network theory, along with several invited reviews, book chapters, commentaries etc. The citation indices of several of these publications were among the highest in immunology at the time, and certainly higher than that of most of my published work on other subjects before or after. I was invited as a main speaker to nearly all international

Autobiographical note

conferences in immunology. I enjoyed the life of an internationally recognized scientist and a prominent professional career in the German academic system.

How can such a thing happen? How is it possible that hundreds of scientists engage in work, over periods of more than a decade, that thereafter gets disposed as meaningless? What were my own motivations and perhaps illusions that made me a follower? While error is normal in experimental work, it usually happens to individual scientists or perhaps members of a laboratory, is of short duration and eliminated by lack of reproduction. Scientific fraud does happen in the scientific community, and may have occasionally occurred in idiotypic network research as well. However, collective fraud is not conceivable as the origin of the network paradigm, and in this essay I will not be concerned with the minor role it may have had.

In search of explanations I came across a bunch of scientific literature which, unknown to most experimentalists, aims at understanding the generation of knowledge from scientific processes, a science known as epistemology (from Greek: επιστημη = knowledge), recently more often referred to as science theory or science studies. Like most experimental scientists I was vaguely familiar with the major works of the famous science philosophers Karl Popper[8] and Thomas Kuhn[9], without having read their writings in detail. However, I had no knowledge whatsoever about a sizable cohort of publically less known science theoreticians, science historians and sociologists, and their various approaches to analyze what experimental scientists do and achieve in the laboratory. Indeed, they seem to look at us in a way similar to how anthropologists look at newly discovered human tribes or behavioral scientists at their laboratory animals. The field of epistemology has been and continues to be highly active, with as many facets and diverse points of view as there are contributors. We experimental scientists are of the opinion that what we do in the laboratory is finding the truth, generating solid knowledge, uncovering facts about nature and how it works. Far from that, according to our epistemologists we may generate measurements, theories, or paradigms, but whether our work generates facts or reveals the truth about nature are quite different questions altogether.

Writing this book had initially been an attempt to understand what made me participate in and contribute to what now seems to be a major scientific vagary. In the course of reading and writing my intent shifted to learning about the philosophical constraints that govern experimental science in a more general way. I learned that much of the past and present science philosophy has been contributed by scholars trained as physicists, and that their thoughts do not always seem suitable for dealing with the epistemological problems associated with biology. By comparing the network vagary with more successful explorations in the life sciences I arrived at a number of conclusions as to how scientific notions turn, or may not turn, into scientific facts in biology. Some of this has not been considered by others, at least not in the same way, perhaps because only few of the science philosophers have been biologists.

I begin this book by a brief summary of what appeared to me the most enlightening concepts and conclusions in modern epistemology. I must confess, however, that epistemology is a huge field and much of it I am unable to fully appreciate. My selection is thus biased by what seemed sensible reasoning to me, a lay philosopher. Some of the difficulty is due to the fact that epistemological literature uses a unique language and terminology, which can be quite cumbersome to read for an experimental biologist who is used to a quite different sort of language. Because this book is written for my colleagues, I try to convey my thoughts in the language we use to communicate our results and conclusions in scientific papers. I am sure that this is not doing full justice to the complexity of epistemology but, anyway, this is the only way I can write. In the second part I shall reconstruct the rise and fall of the network theory. Finally, in the third part, I will examine the network paradigm in the light of some of the existing concepts in epistemology, adding some of my own thoughts on the generation of knowledge in the life sciences. The title: "The Network Collective" is an adaptation of Ludwik Fleck's term "thought collective", meaning a group of scientists who share a certain "thought style", a manner of thinking and reasoning in science that is subject to change over time[10].

References

1   Eichmann K (2005) *Köhler's Invention*. Birkhäuser Verlag, Basel
2   Burnet FM (1959) *The clonal selection theory of aquired immunity*. Cambridge University Press, Cambridge
3   Brait M, Vansanten G, Van Acker A, De Trez C, Luko CM, Wuilmart C, Leo O, Miller R, Riblet R, Urbain J (2002) Positive selection of B cell repertoire, idiotypic networks and immunological memory. In: Zanetti M, Capra JD (eds) *The Antibodies*, Vol 7. Taylor and Francis, 1–27
4   Marshall E (1996) Disputed results now just a footnote. *Science* 273: 174
5   Cohn M (1994) The wisdom of insight. *Ann Rev Immunol* 12: 1
6   Jerne NK (1974) Towards a network theory of the immune system. *Ann Immunol (Inst. Pasteur)* 125C: 373
7   Jerne NK (1976) The immune system: a web of V-domains. *Harvey Lectures* 70: 93
8   Popper KR (1959) *The Logic of Scientific Discovery*. Hutchinson, London
9   Kuhn TS (1962) *The Structure of Scientific Revolutions*. The University of Chicago Press, Chicago/London
10 Fleck L (1979) *Genesis and Development of a Scientific Fact*. Trenn TJ, Merton RK (eds) The University of Chicago Press, Chicago/London (Original in German: Fleck L (1935) *Entstehung und Entwicklung einer wissensachaftlichen Tatsache; Einführung in die Lehre vom Denkstil und Denkkollektiv*. Benno Schwabe und Co. Velagsbuchhandlung, Basel)

Part I

Scientific Knowledge, Delusive or Deductive

Part 1

Chapter 2

Realism, constructivism, and the naiveté of the experimental scientist

Given any pair of properties that have previously always occurred together in our experience, it is possible to construct an infinitely large variety of contrary theories, all of which are compatible with the inductive rule: "If, for a large body of instances, the ratio of successful instances of a hypothesis is very high compared to its failures, then assume that the hypothesis will continue to enjoy high success in the future".

L. Laudan (1996)
Beyond Positivism and Relativism

The theory of science is as old as science itself, its beginnings going back to Aristotle and possibly further. Its history includes names as prominent as Descartes, Diderot, Kant, Fichte, Hegel, Wittgenstein, among others[1]. It is also just as heterogeneous as science itself, as it has produced a large number of different theories which can only roughly be divided into realistic and constructivistic/relativistic ones. Realists believe in scientific truths and therefore insist that scientific objects exist in reality. Constructivists, in contrast, think that scientific objects, for example quarks, are not real but constructed fictions which are useful to formulate scientific theories which may be empirically adequate to describe experimental results, but not the real world. Each of the two groups of theories comes in a number of variants which attend to different aspects not all of which appear equally plausible to the experimental scientist and lay philosopher. I therefore restrict myself to discussing schools of thought which I conceive as major, and as of some relevance to the life sciences.

In the framework of realism, positivism is the term by which philosophers of science refer to the tendency to place science, particularly the natural sciences, in the leading cultural position, and to recommend scientific methodology as a model for all theoretical and practical human activities. Historically, positivism is quite ancient, the term having been coined by Auguste Compte in the 18th century[2,3]. Compte distinguished three stages of progress in human development, which he meant to occur similarly in individual ontogeny and in the phylogeny of mankind: A theological stage, in which there is total freedom of spontaneous imagination or fiction with no requirement or even admittance of proof, corresponds to an infant's way of thinking. This is followed by the adolescent metaphysical stage in which thinking is guided by ideas or entities that spring from the mind solely by abstract reasoning. The adult – positive – stage humbly relies on conclusions from exact data derived by observation, analysis, and proof. This first notion of positivism was taken up and developed in different directions by the French philosophers E. Littré and P. Lafitte, and by H. Spencer and J.S. Mills in England, which gave different weight to humanistic sciences, such as sociology, economics, religion, morale, etc., in their philosophical systems. Later, in the 19th century the physicists E. Mach and H. Poincaré overthrew these viewpoints, by denying these humanities the objectivity and rigorous methodology of the natural sciences, raising particularly physics to the place of pride. Mach, who commanded a broad spectrum of scientific disciplines including, in addition to physics, psychology, behavioral sciences, and physiology, attributed great importance to observation as the prime source of human cognition, which in his view was essentially empirical. Observations could be made directly by sensorial perception as well as indirectly by measurements using scientific instruments[4].

Neo-positivism, finally, is attributed to the deliberations of the Vienna Circle and the Berlin School of Scientific Philosophy in the late 19th and early 20th centuries. Neo-positivism stressed the importance of empiricism as the center of all cognitive activities, unlike Kant's belief that causality, as exemplified by Newton's laws of mechanics, is an *a priori* principle that cannot be empirically ascertained or refuted[5]. In Berlin, the logical empiricist Hans Reichenbach postulated that human thinking is guided by the conviction that nature functions according to a set of hidden laws which are not empirically obvious. In our activities we have no choice but to try and unravel these laws using logic, objective justification and mathematical equations as methodology[6]. In Vienna, a unique community of intellectuals including, among others, the mathematician Hans Hahn, the philosopher Rudolf Carnap, the universalist Otto Neurath, and the physicist Moritz Schlick, proposed a rigorous scientific philosophy that concentrated on epistemologic, logic, and linguistic analyses[7]. The source of cognition is purely empirical, the method of analysis is logic. These attributes, inherent to the natural sciences that serve as a model, were to be adopted by all other scientific disciplines including the social sciences and humanities, thus demanding a unified scientific approach. A leading treatise was Schlick's "General Doctrine of Cognition"[8], in which he emphasized the empirical approach, based on objective observation and unbiased judgement, as the only valid source of cognition. Logical judgements of objectively observed empirical facts will reveal the true nature about the world around us, and will successfully guide us to deal with all its challenges. This attitude of the positivists, optimistic as it was, did not hold for very long. Contemporary philosophers reject positivism/realism on two types of grounds: On the one hand, there is now a general consensus that empirical cognition is impossible without preexisting theory; on the other hand, meta-empirical, i.e. social, factors in science are not taken into consideration.

In contrast to realists, constructivists attribute a central role in human cognition to the notion of theory. While constructivism has overthrown realism as a school of thought in epistemology, its roots can be traced back to a contemporary of the founder of positivism, Auguste Compte, the English universalist David Hume[9,10]. Less rigorous with religion than Compte, Hume proposed a mild academic scepticism which denied any possibility of knowledge about the metaphysical or transcendental, which may exist but cannot be ascertained. With science, however, Hume was a lot more sceptical than Compte[11,12]. While Compte believed in science methodology as a general approach to positive knowledge about all subjects of human interest, for Hume this is restricted to geometry and mathematics. Triangles and circles are products of pure logical thinking and independent of their existence anywhere in the world. In contrast, cognition about nature by logical thinking is principally impossible and the widespread conviction that nature functions according to constant rules is not born out of logic. Instead, our perception of reality rather depends on human experience by observation and empirical connections. We get accustomed to connections between facts or events by frequent repetition and submit them to memory, leading to the experience that (say) facts A and B are connected. We then assume that we understand the causality of the connection, but the forces behind such causalities remain obscure. For example, we empirically know that a stone without support (A) falls down to earth (B), we think we understand this as caused by gravity, but we do not know the causes of gravity. On the basis of our memory we then believe that A and B are always connected and will be connected in all possible instances in the future. We have thus deduced a theory about the connection of A and B which is supported by nothing but our past experience. However, according to formal logic there is no reason to postulate that nature functions by constant rules, i.e. there is no *a priori* logical ground for A and B to remain connected at all times. Someone relying on pure logic could with equal right propose the opposite theory, i.e. that no connection exists between A and B, and wait for future instances to make the choice between the opposing theories. Facts thus can be known only on the basis of experience, and our actions are guided by memory rather than by knowledge. However, according to Hume the ability to act according to memory is a "natural instinct" and as such a biologically important and useful human property, irrespective of causalities being real or just imagined.

Contemporary constructivists have drawn attention to the artificial nature of laboratory research[13]. Experiments are performed under conditions which eliminate all environmental influences, natural objects are transformed by detection devices into computer images, graphs on paper, bands on autoradiographs, etc. Using scientific apparatus, natural objects are detected rather than observed. Observation by the naked eye differs from detection by a scientific device, for example an electron microscope, by the degree of immediacy. A large object, such as an elephant, can be observed directly, a very small object, such as a virus, is detected by the electron microscope. Still smaller objects such as elementary particles are detectable merely by the traces they leave by condensation in a vapor chamber. Detection differs from observation because, rather than seeing the object, we see, with the help of some apparatus, an effect of the object on another matter, such as condensation of droplets out of vapor. In an electron microscope we can detect an object only after it has been treated in a variety of ways, such as freeze-etching, embedding, gold coating, and the like. We thus do not see the virus particle, we see the traces that it has left in the media used to prepare the sample. Also simpler devices such as a light microscope require some form of treatment, such as staining with dye, to make objects detectable. Objects that are large but very distant are indirectly detected as well, such as a jet plane high up in the sky leaving a condensation track that we can see, or a star extinguished long ago that we detect by its light reaching us after zillions of years. It is thus very difficult, if not impossible, to draw the line between observation and detection and, anyway, most objects of scientific interest are detected rather than directly observed. The implication of this is that detection requires that we believe in the connection between the object and its traces, just like that between A and B discussed above. Detection of objects thus requires theory, it is theory-laden in the philosopher's jargon. This includes most forms of observation as well, as detection and observation cannot be reliably distinguished.

The conclusion that even single events of scientific detection require theory entails that comprehensive scientific theories, which connect multiple such observations and detections, can be broken down into individual theses or statements. The doctrine of reductionism in epistemology states that such a comprehensive theory can be empirically examined by examining, one by one, the individual theses of which it is composed. Willard van Orman Quine, American philosopher of the 20th century and leading relativist, opposed reductionism by stating that a theory is to be examined as a whole, to be confronted with the "tribunal of experience", because evidence contrary to an individual component of the theory can always be accounted for by making adjustments in some other component of the "web of belief", Quine's term for theory[14]. This statement, termed holism as opposed to reductionism, is known as the Duhem-Quine thesis, acknowledging the French physicist/philosopher Pierre Maurice Marie Duhem. As a natural scientist, Duhem contributed to many subjects including mathematical physics, thermodynamics, hydrodynamics, elasticity, mathematical chemistry, and mechanics. His philosophical vision, however, was a unifying thermodynamic theory to explain the whole of physics and chemistry[15]. An often used example to illustrate the Duhem-Quine thesis refers to Newton's mechanics which, while making impressively precise predictions on the motion of objects on earth as well as of the planets, failed to correctly predict the orbit of Uranus. Newton's followers did not discard Newtonian mechanics as a falsified theory but supported the theory by making the following adjustment or auxiliary hypothesis: They postulated the existence of an additional planet, unknown at the time, that influenced Uranus's orbit. This planet, Neptune, was later indeed discovered. Quine, a radical logician, concluded that any theory can be reconciled with any recalcitrant evidence by making suitable adjustments, so-called "ad hoc hypotheses"[16,17], in our assumptions about nature. Conversely then, a theory can never be unequivocally deduced from the available evidence or, any theory can be deduced from any available evidence. Some relativistic philosophers have concluded from these and related considerations that scientific methodology is principally insufficient, a view which is reflected in the doctrine of "underdetermination of theories"[18]. It says that, given any body of evidence, one can make any number of different or even contradicting hypotheses to fit that evidence, and there are no logical grounds to make a choice among them. Connaisseurs of the field differentiate deductive underdetermination, referring to the deduction of theories from observable evidence, from inductive underdetermination, referring to the tendency of theories to make predictions about instances which are unobservable and/or lie in the future. The doctrine of underdetermination has been formulated by different philosophers in different ways, but all more or less radically denying the validity of scientific methodology[18].

Nevertheless, practicing scientists ignorant of underdetermination have always chosen and continue to choose between competing theories on the basis of, as they like to think, unequivocal evidence. In order to explain theory choice in the face of underdetermination, relativist philosophers have invoked subjective and non-cognitive factors as a basis for theory choice. As suggested by Mary Hesse, scientists are led by "social factors" to choose between theories[19]. David Bloor concluded, using historical examples, that scientists may even change their "system of belief if there is no change whatsoever in the evidential basis"[20]. Paul Feyerabend likes to persuade us that the only difference between successful and unsuccessful theories lies in the promotional talents and powers of their respective advocates[21]. Feyerabend even proposed that scientists should be controlled by civil committees as to what they are allowed to investigate and what conclusions to be drawn from the results[22]. The "scientific nihilism" resulting from the notion of underdetermination clearly poses a serious threat to science.

The probably most radical version of relativism has been formulated by the German science theoretician Kurt Hübner. In his books *Kritik der wis-*

senschaftlichen Vernunft (Critique of Scientific Reason)[23] and *Die Wahrheit des Mythos* (The Truth of Myth)[24] he compared mystic views of the world, in his terminology "mystic ontologies", as they are inherent to religions, with the scientific view, i.e. the "scientific ontology". Arguing that the history of science is a series of unsuccessful or rejected attempts to gain definitive insight into the foundations of reality, he concluded that science like any other ontology ultimately requires a set of *a priori* assumptions which cannot be empirically ascertained. As a result, science has no reason whatsoever to claim superiority over a mystic ontology. From this argument he developed his "first general principle of tolerance: In so far as all ontologies are contingent and fail to have a necessary validity, none can be preferred over the other".

Luckily, practicing scientists choose and attack their problems rather naively. Either they do not know about the doctrine of underdetermination or, those who do would likely declare it as utter nonsense. Certainly in the life sciences but in other natural sciences as well, the procedures of discovery follow different rules of rationale. In the following I will try to illustrate them using as an example the development of the knowledge that the substance, deoxyribonucleic acid (DNA), is the molecular basis of heredity. Today there is no doubt that genetically transmitted traits are encoded in the DNA. The knowledge is so robust that it has developed into a basis not only of scientific but of many practical human activities, from agriculture to biotech industry, from pharma to medicine. I call this knowledge robust because it has passed, and continues to pass, whatever trial of strength it is put to. The critical experiments that established the connection between DNA and heredity were performed between 1930 and 1943 by Oswald Theodore Avery, a physician of Canadian origin who worked as a laboratory scientist at the Rockefeller University in New York[25]. Together with his colleagues Colin MacLeod and Maclyn McCarty, Avery was studying the pathogenesis of pneumonia, then a devastating disease and cause of death for many young people, including soldiers. The motivation for their work was that detailed knowledge of the pathogenesis of pneumonia was the prime prerequisite for developing measures for prevention and treatment, if they were at all possible. It was already common knowledge that pneumonia was caused by infection with a bacterium, termed *Streptococcus pneumoniae* or in short pneumococcus. When grown on agar plates in the laboratory one can observe two forms of pneumococcus, one that produces a capsule, forms smooth colonies on agar, and is pathogenic in rats or mice (S-forms), the other without capsule, forming rough colonies, and non-pathogenic (R-forms). The capsule consists of polysaccharides and protects the S-form from host attack *in vivo*, whereas the R-form is unprotected and therefore killed by the host. A few years earlier the English scientist Frederik Griffith[26] had demonstrated that S-forms contained a substance that could turn harmless R-forms into harmful S-forms, by the following experiment: Injection of living S-forms killed a rat but injection of living R-forms or of heat-killed S-forms did not; a mixture of heat-killed S-forms and living R-forms, injected after a time of coincubation in a test tube, killed the rat, and living S-forms could be recovered from the killed rat. He concluded that a heat-resistant substance released by S-forms had transformed R-forms into S-forms, in such a way that the progeny retained S-form characteristics. Griffith speculated that S-forms released a protein that functions as "pabulum" to enable R-forms to generate the capsular polysaccharide. The phenomenon, termed transformation, went largely unnoticed by the scientific community until it began to fascinate Avery, as a system likely to reveal important information on pneumococcal pathogenesis, if understood in molecular terms. He decided to identify the transforming substance, using state of the art technology of the time. Avery and colleagues began by optimizing the Griffith assay, which had been rather unreliable in its original version, i.e. they developed an *in vitro* transformation assay that worked reliably and that freed the system of many unknown factors potentially influencing Griffith's *in vivo* infection experiments. This took a lot of patience and frustrating effort, as Avery said to his colleague Rollin Hotchkiss: "Many are the times that we were ready to throw the whole thing out of the window". Having achieved this they performed two main types of experiments. In the first, they fractionated extracts of S-forms, which contained at least several dozens of different types of substances, into the four major classes of candidate biomolecules: proteins, DNA, ribonucleic acids (RNA), and polysaccharides. Only DNA but none of the other three fractions had transforming capability. They then reasoned that transformation may be due to traces of another, unknown substance contaminating the DNA preparation. In the second experiment, therefore, they treated the DNA preparation with each of the four different enzymes deoxyribonuclease, destroying DNA, ribonuclease, destroying RNA, protease, destroying proteins, and lipase, destroying lipids. Only deoxyribonuclease but not the other enzymes abrogated the transforming activity. As polysaccharides were excluded owing to the purification procedure, they concluded that DNA was the transforming principle, but added the caveat that "the biological activity of the substance described is not an inherent property of the nucleic acid, but is due to minute amounts of some other substance absorbed to it or so intimately associated with it as to escape detection".

Although they considered the possibility, Avery and colleagues carefully avoided any general claim that DNA was the substance of heredity. Indeed, such a claim would have required to be certain that bacterial transformation was a case of heredity and functioned by the same molecular mechanisms. This was far from certain at the time, although it was strongly suspected, particularly because the transformed bacteria stably transmitted the trait to generations of offspring. As a result, the communal perception of the work was that Avery and coworkers had identified the substance of heredity.

Declaring DNA as the substance of heredity was rather unpopular at the time as most of Avery's colleagues had favored proteins whose composition of chains of 20 different amino acids in any possible sequence allowed for pronounced diversity and therefore was much more in keeping with the high degree of variability expected of the substance of heredity. DNA, in contrast, was known to consist of only 4 different nucleotides and a single sugar, deoxyribose, in relatively constant molar ratios and was therefore considered as much too simple for the two essential properties that the substance of heredity was required to possess: To self-replicate indefinitely and to specify an endless multitude of genetic traits. It thus took several years of additional experimentation to convince the scientific community that Avery was right. Alfred Hershey and Martha Chase demonstrated in 1950, by specific radioactive labeling of either DNA or protein, that a bacterial virus, a phage, transmits its DNA but not its proteins to its progeny[27]. In 1953 James Watson and Francis Crick solved the structure of DNA by X-ray crystallography as double helix of two anti-parallel strands of the four different nucleotides[28]. This structure immediately suggested putative mechanisms of self-replication, as the nucleotides of one strand form complementary pairs with that of the other so that the nucleotide sequences of the two strand determine one another. The precise mechanism of self-replication was then identified in 1958 by Matthew Meselson and Franklin Stahl, who demonstrated that the strands of the parental double helix separate during replication, each strand serving as template for the synthesis of a new strand, resulting in two double strands, each identical to the original one[29]. Marshall Nirenberg and Heinrich Matthei uncovered in 1961 how the amino acid sequences of proteins are derived from the nucleotide sequence of DNA: Proteins are synthesized by an intricate process involving several forms of ribonucleic acid (RNA) as intermediates in such a way that DNA serves as template for RNA which then determines the amino acid sequence of protein[30]. Finally, Crick and Sidney Brenner found that the sequence of three DNA nucleotides, a triplet, determines an RNA triplet, which in turn determines one amino acid in a protein chain[31]. Thus, each of the 20 different amino acids in a protein sequence is encoded by one triplet of nucleotides in the DNA sequence. The genetic code is universal, i.e. the same in all known living organisms.

What are the epistemological criteria of the Avery experiment that generated the knowledge: DNA is the substance of heredity? Positivists would happily interpret the discovery as one based on careful and objective observation of nature, followed by logical judgement, i.e. following scientific methodology by the book is bound to reveal the truth about nature. However, so would relativists find nearly all their reservations confirmed as well. The phenomenon of transformation by mixing different types of bacteria or their extracts, treated in different ways with unknown consequences, could not be more theory-laden. Improving the Griffith assay to become reliable is nothing else but eliminating environmental factors, thus creating a laboratory situation far removed from reality. The conclusions rely on detection by indirect means rather than on direct observation. Moreover, conclusions could only be drawn relative to a "web of beliefs" consisting of multiple connections based on experience rather than on knowledge. Experimentalists use the term "black box" to describe the unknown terrain between the mixing of components at the beginning and the measurable effects at the end of an experiment. As a result, the deduction that DNA is the transforming substance was underdetermined by the observable evidence, one only has to quote Avery himself who considered the possibility that another, unknown substance, associated with DNA, could be responsible. Quine would have formulated that the evidence was compatible with any type of theory, i.e. any substance could be responsible for transformation, or with opposing theories, i.e. DNA is the substance – is not the substance of transformation. The induction that DNA is the substance of heredity is even more severely underdetermined, as it requires the conjecture that bacterial transformation is a paradigma* of heredity in general. So how is it possible that this robust knowledge, today a firm foundation of numerous human activities, originates from this experiment?

A few epistemological conclusions can be deduced from this example about the origin of knowledge in the life sciences: (i) It is obvious that our present knowledge of DNA originated from Avery's experiment but was not at once generated by it. Seventy years earlier Johann Friedrich Miescher had discovered that DNA was a main component of the nucleus[32], and the chemical composition of DNA had been worked out since. Subsequent to Avery the confirmation by many other investigators was required to convince the scientific community that Avery's claim was correct. Avery knew as well as any other biologist that a single experiment, even if it is reproducible in one laboratory, is insufficient to establish a fact. All experimentalists know the relief that springs from the information that their experiment was reproduced in another laboratory. This is due to the black box, the multitude of unknown factors in each experiment, some of which may be special to one's own laboratory. Only by the frequent repetition in laboratories under presumably different unknown conditions can an experimental result retrospectively rise to the status of robust knowledge. Scientific knowledge is thus not made at once, it is made by *retrospective confirmation*. (ii) While Avery concluded that DNA was the principle of pneumococcal transformation, only later interpretations extrapolated Avery's suggestion to the discovery of the substance of heredity. While Avery was certainly aware of the possible implications of his work for heredity, he restricted himself to quoting published arguments of others that transformation might be "described as a genetic mutation" or the

---

* I make a difference between paradigm and paradigma. Paradigm is used in the Kuhnian sense (see chapter 3). Paradigma, from the Greek παραδιγμα, means example.

"inducing substance might be linked to a gene". Moreover, only by later discoveries was Avery's suggestion amplified to make sense in a more general context, i.e. that the structure of DNA was indeed suitable of performing the tasks that the substance of heredity had to be capable of. The status of robust knowledge thus requires *retrospective amplification* of an experimental result. (iii) Just as Griffith's experiment needed confirmation and amplification by Avery, by turning it into a reliable phenomenon and by uncovering its molecular basis, Avery's experiment needed confirmation and amplification by his successors. Robust knowledge is thus a *collective achievement*, although contributions by certain individuals may be pivotal. (iv) What Avery wanted to achieve was to understand the phenomenon of pneumococcal transformation, with a view to increase knowledge of the pathogenesis of pneumonia. The connection between bacterial transformation and heredity was very uncertain at the time, and speculations about a mutational process or about a process resembling induction of cancer by viruses are only two of a number of possibilities that were entertained. Bacterial transformation could thus not be used as an unequivocal paradigm of heredity and, therefore, the discovery of DNA as the substance of heredity was fortuitous. Although there are better examples, the discovery of DNA by studying bacterial transformation exemplifies that knowledge in the life sciences is, more often than not, generated *unintentionally*.

Nevertheless, science can generate robust knowledge. Some philosophers, including those most widely known to the general public, at least partially admitted to this more realistic attitude. In contrast to the relativist point of view, i.e. that any evidence is compatible with any theory, these philosophers developed their theories on the basis that scientific methodology can generate evidence that can indeed be used to put a theory to the test. Some of these philosophies will be discussed in the next chapter.

Chapter 3

Beyond underdeterminism: Popper, Kuhn, et al.

Such a stylized solution, and there is always only one, is called truth.... It is always, or almost always, completely determined within a thought style.... Truth is ... (1) in historical perspective, an event in the history of thought, (2) in its contemporary context, stylized thought constraint.

Ludwik Fleck (1979)
Genesis and Development of a Scientific Fact

The common notion of realism is that scientific progress consists of an asymptotic approximation to the truth. The most influential concepts of how this approximation works have been put forward by Karl Popper und Thomas Kuhn. Popper is probably the science philosopher of the previous century best known to the general public, and the basics of his postulates are familiar to most practicing scientists[1,2]. Not generally known may perhaps be that he wrote his most influential work, *Logik der Forschung*, already in the mid 1930s, still living in Vienna, in German. Much better known and usually quoted is the English translation, *The Logic of Scientific Discovery*, that appeared about 25 years later, in 1959[3]. Popper was trained as a physicist and his philosophy was much inspired by the then recent impact of Einstein's special and general theories of relativity, which had appeared during the first decades of the 20th century[4,5]. Einstein's theories represented scientific revolutions in that they unified and replaced prior theories on space, time, light, and motion, etc., theories that had seemed to be solid dogma for long periods of time. Einsteins's theories not only impressed and convinced most of his colleagues, they became rapidly known in the public and had a pronounced revolutionary influence on the general perception of the world. Indeed, the history of theories in physics appears like a corroboration of the inductive rule of underdetermination: There is no logical reason to be sure that a highly successful theory will continue to be successful in the future. As all theories had so far been replaced by new, more accurate ones, no theory can claim to be true. Popper's way out was to suggest that we accept a theory as correct for as long as it has not been disproved by novel observations. This constraint applies not only to the making of theories but extends to their experimental examination as well. An experiment can never prove a theory true, i.e. verification is *a priori* impossible. Conversely, a theory can readily be proven wrong by experimentation. As Popper viewed the progress of scientific knowledge as a succession of theories, each new one replacing older, less satisfactory ones, he actually literally called upon scientists to try and falsify their own theories.

Scientists working in biology and the life sciences always had a hard time accepting Popper's views as accurately describing their own efforts. For once, theories in biology differ from theories in physics. Most modern physics is based on theories such as quantum mechanics, quantum gravitation, string theory, etc., which attempt – though unsuccessfully – to explain the material nature in its entirety ranging from elementary particles to planets and galaxies, including the universe from its origin. Biologists, in contrast, have learned the hard way that statements of generality rarely hold for long, i.e. that any frequently used biological mechanism has its exceptions somewhere in nature. A well known example is the famous theory on the transfer of sequential biological information, put forward in 1958 under the term "The Central Dogma" by Francis Crick[6,7]. On the basis of then limited evidence Crick postulated that sequential information transfer is unidirectional in biology: It can pass from DNA to DNA (replication), from DNA to RNA (transcription) and from RNA to protein (translation). The reverse directions, transfer among RNAs, or transfer by proteins are forbidden. While the dogma held for more than a decade, in 1971 David Baltimore and Howard Temin found out independently that certain viruses whose genetic information is encoded in RNA possess an enzyme, reverse transcriptase, which transcribes viral RNA into DNA which can then integrate into host DNA and thus replicate in host cells[8,9]. Today we know numerous exceptions to the central dogma, for example some RNA viruses can replicate RNA directly. Even proteins can transfer structural information, albeit conformational in nature, as revealed for prions[10]. But is Crick's hypothesis thereby falsified? It is certainly not, as it continues to be correct for the majority of instances. Indeed, the unidirectional transfer of sequential information is robust knowledge for the vast majority of living organisms. What had been falsified was the use of the term dogma, because a dogma is without exceptions. In his autobiography Crick wrote "I had already used the word hypothesis (in another context) and I wanted to suggest that this new assumption was more central and more powerful. ... As it turned out, the use of the word dogma caused almost more trouble as it was worth..."[11]. Crick's theory is a typical biological theory. Because biological theories are expected to eventually encounter exceptions as research goes on, they are not necessarily falsified by discovering exceptions.

Another, perhaps less known example can be derived from the controversy about the nature of antibody specificity which began in the 1930s and lasted for more than two decades. There were two opposing types of theories, the instructional and the natural/clonal selection theories. Instructionalists proposed that all antibodies had the same amino acid sequence, and that specificity for antigens was acquired by antigen-induced folding of the polypeptide chains. Once folded, the conformation was irreversible and the antibody molecule was permanently instructed to specifically bind the inducing antigen. Instruction was for many years the accepted mechanism to explain antibody specificity and supported by many leading scientists of the time. This began to change when Niels Jerne reported on experiments which showed minute amounts of specific antibodies in the sera of non-immunized animals. As a consequence, he and later Macfarlane Burnet proposed that large numbers of different antibodies are preformed in the body and clonally expressed by antibody-producing cells. In addition, the unique specificity of an antibody was determined by its unique amino acid sequence. The evidence that eventually decided the controversy against instruction and in favor of clonal selection will be briefly discussed in chapter 6 and 16 (see also refs. [12,13]). Suffice it to say here that instruction was against the central dogma, which was proposed around the same time, the late 1950s, as the clonal selection theory. The important point in the present context is that in the 1990s several authors including Cesar Milstein, a vigorous opponent of instruction, reported on antibodies which could indeed change their conformation, and consequently their specificities, upon instruction by an antigen[14]. Again, these exceptional observations did not falsify the clonal selection theory, which continues to belong to the robust knowledge in immunology.

As a consequence, biologists have often rejected and sometimes ridiculed Popper's philosophy. For example, Niels Jerne is quoted to have privately said that Popper "spoke a lot of nonsense, and too much", and written "science is not an accumulation of series of falsifiable propositions which experimentalists try to falsify, but also, and mainly, imagination: a development of concepts and new perspectives that change the outlook and the type of propositions, discussions, and experiments"[15]. Subsequent science philosophers have also doubted the validity of Popper's falsification thesis, including Kuhn (see below) and Quine with the Duhem-Quine holism thesis (chapter 2). Similarly, Imre Lakatos found Poppers falsificationism "naive" and noted that, even in the face of recalcitrant evidence, scientists tend to stick to their underlying convictions, which he termed research programs[16]. Only if scientists come to realize that an alternative research program is more promising, i.e. likely to be more successful, will they abandon the existing one[17]. The notion of research program, as used by Lakatos, thus seems very similar to Kuhn's term paradigm (see below).

Of equal if not greater impact as Popper in present day epistemology has been Thomas Kuhn, who published the first edition of his most influential treatise *The Structure of Scientific Revolutions* in 1962[18]. Subsequent editions have appeared since. Trained as a physicist just like Popper, Kuhn left experimental physics because he eventually found the history of science more fascinating than science itself. In studying the works of historians on discoveries in chemistry and physics, Kuhn identified three general principles of scientific progress: First, it is difficult for historians to identify the exact circumstances and individuals involved in, for example, the discovery of oxygen or the concept of caloric thermodynamics. Such major forward steps in knowledge are mostly based on diverse observations and scattered contributions by many individuals, together setting a framework for the emergence of novel ideas, techniques, and research avenues. Second, science historians seemed to be uncertain about how to make distinctions between myths, beliefs, superstitions, errors etc., on the one hand, and clearcut scientific reasoning on the other. When viewed on the background of the knowledge of the time, old now abandoned scientific concepts did not have obvious flaws and appeared no less sound in scientific procedure and reasoning than contemporary concepts of the same subject. He concluded that, in the intellectual context of their time, old scientific theories may be as valid as their modern successors are now: "Out-of-date scientific theories are not in principle unscientific because they have been discarded".

Most important in Kuhn's view was the third principle of science history: Acquisition of knowledge is discontinuous. Historical analyses led Kuhn to suggest that acquisition of scientific knowledge over time cannot be described as a steady process of accumulation of individual discoveries and contributions. As one type of discontinuity, Kuhn distinguishes the early stages of a scientific discipline from its mature form. In the early phase, scientists beginning to address a novel field of study without traditions utilize their own individual experience and knowledge, gained in other subjects, to tackle the new problem. As a result, different scientists looking at the same phenomena arrive at different conclusions and interpretations, all more or less consistent with the phenomena, but incompatible with each other. The early phase of a science is thus often characterized by controversies and competition between a number of distinct views or opinions, with a strong element of arbitrariness.

Effective research begins only after a community of scientists has overcome arbitrariness and begun to agree on certain common entities about the subject under study, the questions to be asked, and the methodology to be used. Kuhn suggests that in history this was often achieved by famous classic writings such as Aristotle's *Physica*, Ptolemy's *Amalgast*, Newton's *Principia* and *Opticks*, Franklin's *Electricity*, Lavoisier's *Chemistry*, etc. These texts were able to unify a previously controversial subject because they had two essential properties: They were sufficiently novel and fascinating to convert a discordant and unstable lot of bunglers into a stable cohort of professional practitioners; and they were sufficiently open-ended to leave much room for further definition and precision to be gained by subsequent conceptual thinking and practical experimentation. Kuhn termed scientific achievements combining these two properties "paradigms". A science reaches maturity with its first paradigm. The paradigm sets the rules according to which a discipline of science progresses for a period of time. It defines the accepted framework of "the given", the open problems, the questions to be asked, and the methodology used by the scientists working in a particular field.

Himself not entirely happy with the term paradigm, which is derived from the Greek παραδιγμα which means something like "example", Kuhn defined paradigms as "accepted examples of scientific practice – including law, theory, application, and instrumentation together – (that) provide models from which spring particular coherent traditions of scientific research". It follows that a paradigm is quite distinct from a scientific theory. It is a much more global entity in that it includes, in addition to a body of theoretical knowledge accepted as facts, elements of scientific practice, such as application and apparatus, that are usually not part of theories. Moreover, while a useful theory makes exact predictions for the results of defined sets of experiments, a paradigm's constraints are more open-ended, providing an experimentalist with creative opportunities for precision, innovation, and discovery. A paradigm may give rise to theories and may comprise one or more theories.

In modern science history, the paradigms of a particular science at a particular time are documented in the textbooks, laboratory methods compendia, and similar writings, of that time. The study of such materials, together with training in the laboratory, entitles a student to membership in the particular scientific community and to advance to an expert in the field. This training involves not only the learning of sets of facts and rules, but also of a body of research traditions, laboratory skills, and the gain of an intuitive understanding of the subject matter. He thus enters a small esoteric circle of advanced experts, leaving a large exoteric community behind. The existence of a body of knowledge unifying the esoteric circle enables its members to build on top of this knowledge and direct their attention to more specialized problems. Kuhn termed research performed under the guidance of a paradigm "normal science". Normal science is what most scientists are doing most of the time. In normal science the specialized work of the experts, and the conclusions drawn from it, are determined and constrained by the framework and the foundations of the paradigm. Normal science is, on the one hand, innovative as its fills the gaps left open in the paradigm, is often directed at highly esoteric problems, and may go into very much depth and subtle detail. On the other hand, normal science is in some respects comparable to "puzzle-solving", as it obeys the rules and its results confirm the expectations of the paradigm, to some extent they are even predetermined by the paradigm. The results of normal science are published in specialized scientific journals. They are often understandable only to members of the esoteric circle, and remain unintelligible to exoteric communities even if they are of related or overlapping scientific background. The paradigm thus enables a scientific discipline to progress in the accumulation of highly detailed, valuable, and precise knowledge about its subject matter.

Viable paradigms are characterized by long-term stability and robust resistance to challenge. Normal science, while generating a majority of results consistent with the paradigm or refining it, sometimes also yields results inconsistent with the expectations of the paradigm. Unless very compelling, these are often ignored or suppressed[19,20]. This is mostly not a conscious decision of the investigator but an inherent tendency of the human mind which tends to suppress perception of the incongruous[19]. Kuhn quotes in detail a report on a psychological experiment in which human subjects were shown series of playing cards in quick succession and were asked to identify them[21]. Among a majority of normal cards were mixed a few anomalous ones, such as a six of spades in red. The result was that the anomalous cards were rarely identified but mostly recognized as their normal counterparts, i.e. as a black six of spades in the example given. Only after repeated exposure some subjects would begin to recognize the faulty cards as anomalous. According to Kuhn, this experiment provides a "cogent schema for scientific discovery. In science, as in the playing card experiment, novelty emerges only with difficulty, manifested by resistance against a background provided by expectation".

Awareness of anomaly is what starts a change in paradigm, according to Kuhn the major source of discontinuity in the accumulation of scientific knowledge. Kuhn gives a number of compelling historical examples in physics in which novel observations caused breakdown of old paradigms. They include, among others, the changes from geocentrism to heliocentrism, from phlogiston to oxygen, from corpuscles to waves, or from Newtonian to Einsteinian mechanics. The impact of scientific revolutions of such magnitude is self evident and need not be elaborated on. However, paradigms may concern smaller areas in science as well, and paradigm shifts may be noticed by scientists only, remaining unknown to laity. Indeed, in a mature science usually several compatible and partially overlapping paradigms coexist, and a particular paradigm change may not concern all of them at a given time. Several such small paradigms and their shifts are described for the field of immunology in chapter 6.

Anomaly is recognized when a major scientific discovery, or a series of repetitive observations do not fit with the expectations of the preexisting paradigm, in an obvious way that no longer permits to ignore them. Precise quantitative measurements often have a critical role here. The immediate consequence of the awareness of anomaly is a period of pronounced professional insecurity, generated by the persistent failure of the puzzles to come out as they should. In Kuhn's terminology, this period is termed crisis, and this is where theories have their place in science. Some theories are put forward to rescue the paradigm, suggesting technical adjustments that attempt to reconcile the anomalies with the existing rules. Such adjustments often contain complicated quantitative predictions, more rapidly resulting in unpleasant complexity than generating satisfactory solutions. Other, more successful theories generate new paradigms, they cause the paradigm shift that eventually terminates the crisis. New interpretations of nature that initiate a paradigm shift, based on discoveries or theories, often emerge in the brains of one or a few individuals, in many cases scientists who are either young or newcomers to the field.

A new paradigm changes the scientist's view of the world, although nothing much may have really changed. Kuhn compares this to the shift in Gestalt-vision examined in psychological tests in which a proband looks at an equivocal drawing that can be recognized, for example, either as duck or as rabbit. Another comparison is the transformation of vision that occurs when a proband is provided with goggles fitted with inverting lenses. After a period of disorientation and confusion seeing the world upside-down, the visual field flips over and is again seen by the proband as upside-up. In both examples, the observed reality remains unaltered, but Gestalt-vision has been dramatically transformed. In a similar way, a scientist's perception of the subject matter is transformed by a paradigm shift. Old measurements and manipulations become irrelevant and are replaced by new ones. New apparatus and methodology are introduced and generate new measurements for interpretation. Often, novel technical terms and language are being adopted. Most importantly, the crisis has been resolved as those observations that had initiated the crisis make sense in the context of the new paradigm. The scientific revolution is over, new textbooks can be written without inherent inconsistencies.

What is achieved by a paradigm shift in terms of scientific progress? During crisis it was eventually realized that the old paradigm was unable to deal with the arising set of problems and thus cannot be maintained as a useful framework for future scientific practice. The new paradigm is able to solve these problems and is thus adopted as a valid and stable guiding system for the future. However, adoption of a new paradigm causes both gain and loss of scientific knowledge. While new paradigms destroy their predecessors, they nevertheless usually maintain some successful solutions of older problems as an integral part. In this way, some of the existing traditions remain valid, and the paradigm change appears as an improvement and an augmentation of the body of knowledge in the field. However, while the scientific community believes to have increased its body of knowledge, no new paradigm can completely maintain all achievements of the past. Some old solutions, including some that had been successful, will be abandoned in the process of revolution. Moreover, while there is mostly increase in detail and precision, there may be loss in breadth and in the ability to communicate with other sciences. It is in this context that the question arises whether progress is an inherent asset of scientific revolutions and a necessary consequence. In other words, is the series of scientific revolutions a unidirectional pathway to the truth?

Kuhn views science as an evolutionary process from primitive beginnings whose successive stages are characterized by a net gain in detailed and refined understanding of nature. In most cases, new paradigms describe nature more accurately than previous ones. However, scientific evolution takes place *from* previous stages rather than *towards* any sort of defined goal, which one would have to invoke if science evolved constantly nearer to understanding the truth about nature. The novel observations that cause crisis and subsequent paradigm change, rather than arising in directed expeditions into unknown territory coordinated by a masterplan, are often generated by unintended explorations resulting in unexpected solutions of experiments or concepts. Kuhn stresses the analogy of the evolution of science with Darwin's theory on the evolution of species by natural selection. In this theory, the sole principles of evolution are the collection of existing species, their randomly acquired genetic variations, and the selective forces of the environment. Darwin did not see any evidence for a directed evolution towards an intended goal, such as the creation of *Homo sapiens*. Similarly, the evolution of science is determined by the existing set of paradigms, their variations accumulated by random scientific explorations of nature, and the selection by their problem solving ability. Kuhn thus concludes that "we may have to relinquish the notion that changes of paradigm carry scientists … closer and closer to the thruth". However, as in evolution *Homo sapiens* happened to arise even though there was no masterplan, truth about nature may eventually be revealed in science.

Kuhn's and Popper's philosophies are to some extent similar as the former viewed science as a succession of paradigms while the latter did so as a succession of theories. For both, science consists in changing concepts of the understanding of nature that are characterized by increasing accuracy. Clearly, with respect to the notion of successive concepts Kuhn's philosophy has used Popper's logic of science as a foundation to build on. However, as it seems to be customary among the philosophers of science to vehemently attack each other's principles, Kuhn distanced himself from Popper in a most decisive way, in that he vigorously rejected Popper's falsification dogma: "…it is just the incompleteness and imperfection of the data-theory fit that, at any one time, define many of the puzzles that characterize normal science. If any and every failure to fit were ground for theory rejection, all theories ought to be rejected at all times". According to Kuhn, during phases of normal science scientists attempt nothing like falsification of the existing paradigm. Rather, in their puzzle solving attempts aiming at extension, enlargement, deepening, and precision, they faithfully obey the paradigm's rules and authority. The observations giving rise to crisis more often than not result from explorations into unknown terrain with unintended outcome. Anomalies thus discovered are often disregarded until frequent repetition, mostly by more than one group of investigators, makes their continuing suppression untenable. By falsifying certain theories of the existing paradigm, such experimental observations verify others that then cause the paradigm shift. According to Kuhn, falsification and verification are thus often unseparable consequences of scientific experimentation.

More recent philosophers have also seen the similarities between Popper and Kuhn and have criticized them collectively, noting that both believed in science as a succession of changing modes of thinking, be they termed theories or paradigms. In doing so, they display either a negative or a positive attitude to science. To the scepticists belongs Ian Hacking[22], who declared Popper's and Kuhn's instability in science a "myth", owing to the uniqueness of Einstein's revolution which they naively took as a general example. At most other times – according to Hacking – the theories coming out of laboratory science confirm the phenomena produced in the laboratory by instruments engineered by the scientists to produce such phenomena. Science was thus characterized by a tendency to self-perpetuate or self-vindicate, resulting in "stability" rather than in progress through revolutions or refutations. A more positive attitude towards science can be noted with Larry Laudan[23], who places Kuhn among the strong relativists by (over-)emphasizing Kuhn's arguments on the resistance of paradigms against recalcitrant evidence due to "individual and subjective criteria" causing scientists to reject unexpected discoveries. In his book *Beyond Positivism and Relativism* Laudan critically analyzes all versions of underdeterminism and concludes that "the relativist critique of epistemology and methodology, insofar as it is based on arguments from underdeterminism, has produced much heat but no light whatsoever". Instead, Laudan believes in the existence of unequivocal experimental evidence and gives as an example the historical controversy between Newtonian celestial mechanics, which predicted the shape of the earth as bulging out along the equator and flattened at the poles due to rotation on the axis through the poles, as opposed to Cartesian cosmogony that postulated a perfect round shape or an elongation along the polar axis. Careful and elaborate measurements performed by several expeditions to Peru and Lapland, organized by the Paris Academy of Sciences in the early 18th century, proved beyond doubt that the Newtonian predictions were correct. Laudan formulated as a methodological rule: "When two rival theories, T1 and T2, make conflicting predictions which can be tested in a manner which presupposes neither T1 nor T2, then one should accept whichever theory makes the correct predictions and reject its rival". At a glance this may seem rather trivial, but after having read a lot of relativism and underdeterminism a statement of such unequivocal clarity is certainly comforting.

Chapter 4

The anthropology of science:
Ludwik Fleck et al.

> *...those who consider social dependence a necessary evil and an unfortunate human inadequacy which ought to be overcome fail to realize that without social conditioning no cognition is even possible.*
>
> From Ludwik Fleck (1979)
> *Genesis and Development of a Scientific Fact*

The philosophers discussed in this chapter share the common emphasis of the notion that scientific knowledge is acquired by the undertakings of human beings who act as part of social environments such as academic colleges or laboratory groups. Although Ludwik Fleck was ignored for a long time, most contemporary philosophers concerned with the social conditioning of cognition now agree that his contributions are among the most pioneering and pivotal in this field. Fleck, a Polish medical microbiologist, wrote his main work in German. It was published in 1935, the same year as Popper's *Logik der Forschung*, by a Swiss publisher in Basel, under the title *Entstehung und Entwicklung einer Wissenschaftlichen Tatsache* with the subtitle *Einführung in die Lehre vom Denkstil und Denkkollektiv*[1]. Fleck's book remained virtually unnoticed and it took until 1979 for an English version to appear, entitled *Genesis and Development of a Scientific Fact*, but without the subtitle, which could have been translated as *Introduction to the Doctrine of Thought-Style and Thought-Collective*[2]. It was edited and translated by F. Bradley, T.J. Trenn and R.K. Merton, and includes a foreword by Thomas Kuhn. Indeed, it was Thomas Kuhn who had first discovered Fleck for the western philosophical world. In his foreword Kuhn writes that, when he was about to switch from physicist to philosopher around 1949, he came across a quotation of Fleck when reading Hans Reichenbach's *Experience and Prediction*[3]. Reichenbach used Fleck's illustrations of the changing perception of human anatomy in history to make his positivistic point about how subjective intuition should not have a place in science. Kuhn, on the contrary, immediately realized that there had been somebody, namely Fleck, who thought in terms of changing scientific concepts in similar ways as he, Kuhn, himself. In his foreword to his own book *The Structure of Scientific Revolutions*[4] Kuhn mentions Fleck's monograph as an "essay that anticipates many of my own ideas", and "Fleck's work made me realize that those ideas might require to be set in the sociology of the scientific community". Perhaps because Kuhn failed to quote Fleck specifically anywhere else in his book, Fleck's work still went largely unnoticed until 1977 when W. Baldamus, a German refugee and retired professor of philosophy in Birmingham (U.K.), stimulated his student T. Schnelle to systematically research Fleck's biography and bibliography. The results of Schnelle's efforts are summarized, together with other materials on Fleck, in Cohen and Schnelle (1986)[5].

Fleck's biography[5] needs a brief account here as many of his thoughts become more readily assessable on a biographical background. In particular, Fleck belonged to the rare species of life scientist who were active in biomedical research and in parallel were deeply interested in science theory. He was born in 1896 to Jewish parents in the Polish town Lvov, which until Polish independence in 1918 belonged to the multinational state Austria-Hungaria. He spoke German as well as his mother tongue Polish, and the intellectual environment in his early years was equally influenced by the empire's cultural center, Vienna, and the Polish intellectual community in Lvov and beyond. He was thus exposed to the deliberations of the Vienna Circle as well as many Polish intellectuals. Among others, he had personal contact with the philosophers K. Twardowsky, L. Chwistek, K. Ajdukiewicz, T. Kotarbinski, W. Tatarkiewicz, known as the Lvov-Warsaw School[6]. He studied medicine at Lvov University and thereafter joined the typhus specialist Rudolph Weigl as an assistant. Because of antisemitic policies in the University Fleck had to quit his position in 1923 and worked first as a laboratory bacteriologist at the Lvov General Hospital and later in his own private bacteriological laboratory. In addition to routine work in diagnoses of typhus, syphilis, and tuberculosis, he was engaged in research on serology, skin reactions, and leukocytes, subjects on which he published 39 original papers between 1922 and 1939 in prominent German, English, and French scientific journals. In parallel, between 1927 and 1936, he published four philosophical articles[7], and from about 1932 he worked on his seminal monograph, which appeared in 1935[1]. With the Nazi occupation of Lvov in 1941 he and his wife were deported to the Jewish ghetto where he worked in the hospital on the development of a typhus vaccine under the most primitive of conditions. He immunized first himself and his family before using it on others, with most of the immunized surviving the inevitable infection, the mortality of which under the conditions in the ghetto was close to 100%. The German occupants became interested and brought Fleck in 1943, after a short stay in Auschwitz, to the Buchenwald concentration camp. There he met Eugen Kogon who after the liberation was charged with documenting on the concentration camps and who described Fleck's activities in detail in his book *Der SS Staat*[8]. Fleck had been appointed leader of a group of largely untrained personnel and was charged with the production of a typhus vaccine for immunization of German soldiers. The procedure used was flawed and Fleck later admitted to having consciously supervised the production of an inactive vaccine which did not contain typhus antigens and did not confer protection against typhus to the about 20,000 German soldiers immunized with it. He got away with it and in 1946 Fleck wrote an article, published under the title "Problems of the Science of Science"[9], in which he analyzed, using various historical examples as well as his own experience, the social forces that cause a scientific community to perpetuate a scientific error which is not corrected by mutual silent agreement even though it has been recognized as such. After Buchenwald was liberated by the American army in 1945, Fleck continued his academic career as a research scientist in Poland and became full professor in 1950. He became well known in Poland and abroad for his research on leukocytes, supervised a big laboratory, and collected many honors. Between 1946 and 1957 he published another 87 research articles, but only one – his last – philosophical paper in 1947[10]. In 1957 he and his wife emigrated to Israel were he died of a heart attack in 1960.

In his monograph Fleck developed his epistemological doctrine of thought-style and thought-collective on the basis of the development of the Wassermann reaction, which has been the main diagnostic test for syphilis over many decades, and which was performed in Fleck's laboratory every day. Fleck begins with an account of the changing historical perceptions of how to define syphilis as a disease, of which he distinguishes four phases: (i) a mystic-ethical, (ii) an empiric-therapeutical, (iii) an experimental-pathogenic, and (iv) an etiological definition. In the 15th century mystical-ethical criteria were used to delineate a complex group of diseases which were contracted as the result of promiscuous sexual activities. These venereal diseases, which might have included chancre, gonorrhea, lymphogranuloma inguinale and others, in addition to syphilis, increased particularly in times of war or natural catastrophes, and were thus viewed as punishments inflicted by God for sinful behavior. Astrology had a dominant role in the prediction of catastrophes and its use as a guideline for human activities had a pivotal role in the public perception and professional management of the "carnal plague". These social and historical traditions in the delineation of syphilis were very resistant to challenge and remained valid for more than 400 years, until at the end of the 19th century a second concept slowly took over. It was based on the therapeutic effect of mercury in the treatment of skin ailments, which served as criteria for the delineation of a group of diseases as syphilis, also termed "Scabies grossa" at the time. This concept of syphilis was thus based, in the physicians' jargon, on a definition "ex juvantibus" (by that which helps), in modern scientific medicine considered an improper way of diagnosis. Consequently, syphilis was still not clearly differentiated from chancre, gonorrhea, and many other venereal conditions affecting the skin. The mercury definition, unsatisfactory as it was, became obsolete when physicians began to rely to a greater extent on observations of disease progression, pathology, pathogenesis, and contagiosity. The relationships of the long term neurological manifestations such as progressive paralysis and tabes dorsalis to syphilis were recognized. Secretions and skin materials were tested for whether they were able to transfer disease to healthy individuals, leading to the distinction of syphilis from non-syphilitic conditions. Fleck's fourth – the etiological – stage began with the discovery of the etiologic agent, the spirochete *Treponema pallidum*, in syphilitic material in 1905 by Erich Hoffmann and Fritz Schaudinn, in the context of many similar discoveries for other infections in the late 19th and early 20th century. It is interesting to note, however, that Fleck was less than convinced that the etiologic concept was the final one. In his view it just happened to be the most recent thought-style, transient as the previous ones had been, likely to be replaced by changing concepts in the future.

It was in the experimental-pathogenic phase that the idea of "contaminated blood", "syphilitic blood", or "bad body fluids" first entered the thinking, which was to become, according to Fleck, one of several important factors that led to the development of the Wassermann reaction. A number of additional factors, of scientific and social origin also played a role: The science of serology had begun to develop, as it was realized that contagious diseases cause "immune bodies" – later dubbed antibodies – to appear in the blood that specifically reacted with the infectious agent or its products, as demonstrable in various assays. *T. pallidum* had been identified as causative agent. The French researchers Bordet and Gengou had developed the complement fixation assay in which a standardized hemolytic system was used as an indicator to visualize otherwise invisible antibody/antigen interactions by the consumption of complement. The assay was particularly elegant as it could be used to detect antigen/antibody interactions in which neither the antigen nor the antibody had been previously defined. Adolf von Wassermann, a student of Paul Ehrlich and working at Berlin's Robert-Koch-Institute, and his coworkers had reported on the use of the complement fixation assay to demonstrate anti-tuberculin antibodies in patient material. Although these results had been questionable and did not have consequences for the diagnosis of tuberculosis, they served as starting point for the Wassermann reaction. Another starting point was a feeling in the German ministry of science and education that French scientists were making more important contributions to medicine than Germans, so that something should be done to increase German visibility and reputation. As a result, Wassermann was summoned to the ministry and charged with the development of a diagnostic test for syphilis.

The first version of the test was aimed at the detection of syphilitic antigens in patient tissue extracts using antisera produced by infecting monkeys. These antisera were mixed with minced tissue from syphilitic patients and a positive reaction was revealed by complement consumption in the hemolytic test system. In a second version, the test was adapted to search for antibodies in the sera of syphilis patients, which were mixed with tissue from a proven Wassermann-positive patient and tested for complement fixation. The published initial results showed that the assay for syphilitic antigens was positive in about 90% of the samples from infected individuals, whereas the assay for antibodies was positive in only about 20% of syphilitic sera. As a result, the initial publication emphasized the value of the assay for detection of syphilitic antigen in patient tissues. Not much later it turned out, however, that samples from healthy individuals also often gave positive reactions in the assay for syphilitic antigen, so that this first version of the test was no longer pursued. Instead, the test for antibodies became the subject of interest. While initially giving merely 20% positive reactions, subsequent publications reported a frequency of 70–90% positive reactions with patient sera, with less than 10% false positive results with normal sera. This increase was a remarkable development for Fleck, as nowhere in the literature could he find technical improvements of the procedure which would explain the improved results. Moreover, only one year later several laboratories reported simultaneously that instead of extracts of syphilitic tissue, positive reactions could as well be achieved when normal tissue extracts were used as test antigen. This was a first hint that what was detected by the Wassermann reaction was not, as had been thought, antibodies to syphilitic antigens. Worse than that, it was shown that intentionally elicited antibodies to the spirochete were unable to elicit a positive Wassermann reaction. The initial concept of Wassermann and coworkers, that is to diagnose syphilis by a specific reaction of anti-syphilis antibodies with spirochete antigens, was thus essentially disproved. The assay required neither specific antibodies nor specific antigen to yield a positive result. Fleck: "It is possible to obtain a positive Wassermann reaction from a normal blood sample and a negative one from a syphilitic sample without any major technical errors." Nevertheless, the Wassermann reaction proved itself a valid diagnostic test for syphilis with enormous practical value in medicine, although it required intensive training and expertise in licensed laboratories to perform it in a reliable way. Fleck: "…the Wassermann reaction merely proves the special change in syphilitic blood and we still do not know more…" and "Wassermann and coworkers experienced the same fate as Columbus, they searched for India and were convinced to have been on the correct route, but they found America".

Fleck sees the development of the Wassermann reaction as an example for how a "correct finding can arise from false assumptions and irreproducible initial experiments". It must have been this paradox that prompted Fleck to develop his far-sighted epistemological theory. There are several pivotal categories of reasoning which he introduced to make the linkage of the Wassermann reaction to syphilis understandable, all of which proved of striking general applicability in the theory of cognition. His main categories are that of "thought-style" and "thought-collective". Thought-style refers to a way of thinking about a scientific subject, such as syphilis, or group of subjects, such as diseases in general. Thought-styles change in history, as exemplified by the changing perception of syphilis over the centuries. A thought-style is the interpreted knowledge on a given subject at a given time. It consists of the empirical facts accumulated in a given field of scientific activity, shaped and constrained by non-empirical traditions and subjective fictions of the mind. Thought-styles are shared by the members of a thought-collective, a group of individuals which form the esoteric community of experts in a given field. In Fleck's own words: "…we define thought-collective as a community of persons mutually exchanging ideas or maintaining intellectual interaction, (which) … provides the special carrier for the historical development of any field of thought, as well as for the given stock of knowledge and level of culture. This we have designated thought-style". It is through these intellectual interactions that thought-styles are stabilized but also change and develop. An important factor in such interactions is the inaccuracy of language and the inability to exactly convert one's thoughts into spoken or written words. The remarks of one person are never understood by another exactly as meant. Exchanges taking place within the thought-collective are often understood more correctly and, as they are more frequent than outside communications, serve to stabilize a thought-style. A thought-style is thus quite resistant to change, even in the face of adverse evidence. Interactions between the esoteric community and a larger exoteric laity, including members of unrelated thought-collectives, are frequently misunderstood and tend to cause changes and developments. Multiple such exchanges of opinions on a subject thus eventually lead to changes and evolution of thought-styles. What is perceived as fact by members of a thought collective is thus a transient stylized solution of a problem, a thought constraint. Therefore: "It is altogether unwise to proclaim any such stylized viewpoint, acknowledged and used to advantage by an entire thought-collective, as 'truth or error'. Some views advanced knowledge and gave satisfaction. These were overtaken not because they were wrong but because thought develops. Nor will our opinions last forever, because there is probably no end to the possible development of knowledge as there is probably no limit to the development of other biological forms". The similarities to Kuhn's changing paradigms are striking here, up to the analogy with the evolution of species.

Fleck concludes that truth or fact are categories conditioned by thought-collectives and are thus subject to change. This in turn means that knowledge, in addition to an empirical stock of invariables, always includes presuppositions or stipulations, which Fleck terms "active linkages". Such linkages are historically and socially conditioned elements of knowledge which are actively proposed by the thought-collective as its system of opinions. While members of the thought collective may accept them as truth, they are in reality arbitrary and exchangeable. In contrast, "passive linkages" are consequences that follow from active linkages as a matter of logical necessity. Passive linkages are solid elements of knowledge and constrain the freedom of thinking and scientific activities in a thought collective. Nevertheless, as passive linkages follow from active linkages, all knowledge is linked to thought-style and thus conditional, according to Fleck.

Another category introduced by Fleck is that of "pre-idea". While thought-styles change, they do not do so completely so that some elements are carried on. Pre-ideas are remnants of ancient thought-styles that have not been consciously abandoned and continue to shape the present thought-style, often in an unconscious or unspoken way. Pre-ideas, born within ancient thought-styles, often are somewhat mystic fictions which, when carried on, assume a different meaning, adjusted to the new thought-style. A typical example of a pre-idea is the ancient concept of "syphilitic blood" which had been pursued by many researchers before Wassermann in attempts to demonstrate changes in agglutination, precipitation, hemolysis, flocculation, etc., of serum or blood from syphilis patients, but all without lasting success. Finally, the pre-idea of syphilitic blood became reality, endowed with a new meaning, in the form of the Wassermann reaction, although the initial scientific presuppositions in its development had failed. Nevertheless, "this discovery (of the Wassermann reaction) initiated some very important lines of research; and without much exaggeration it can be considered an epoch-making achievement". It was the beginning of a novel thought-style, carried forward by a new thought-collective.

In Flecks's view, the acquisition of scientific knowledge is a multifactorial process. For the Wassermann reaction it was a combination of political, logistic, medical, and scientific factors, involving the contributions of many individuals, that made the invention possible. There was the political interest, i.e. the desire of the German ministry of science to gain international prestige, that started the whole project. There was the pre-idea of syphilitic blood, coming together with the advent of serology. There was the miraculous improvement of the hit ratio of the antibody test for which there is no accountable scientific reason, likely the result of numerous minute technical adjustments. Many of these multiple circumstances remained unnoticed or were quickly forgotten. Fleck was intrigued by the fact that the leading experimental coworker of Wassermann, one Dr. Bruck, in an article written 18 years later recalled the historical development of the Wassermann reaction incorrectly: He claimed the success rate of the assay to have been satisfactory from the beginning, which was clearly not the case as documented in the published account of the initial experiments. Fleck concludes that the individual scientist is unable to retrospectively trace the multiple detours and errors in the development of a scientific project. Rather, in retrospect the way to success is idealized as a straight progression to the intended goal.

Fleck: "It is difficult, if not impossible, to give an accurate historical account of a scientific discipline".

Of the four epistemologic categories introduced in chapter 2 in the context of Avery's experiment (retrospective confirmation, retrospective amplification, collective achievement, unintentionality), the latter two are particularly applicable to the Wassermann reaction. It was the result of a collective effort that permitted the increase in the ratio of positive instances of the Wassermann reaction from 20% to over 80% over time, in the absence of definable technical improvements. Moreover, the issue of unintentionality takes a particularly absurd twist in this case. The intention to generate a test for syphilis was indeed achieved. However, the scientific assumptions on which the test was designed were false, and its real biological basis remains elusive until to date.

However lucidly Fleck's philosophy reflected the situation of syphilis research at his time, it seems to neglect the fact that a thought-style may eventually stabilize and assume the status of robust knowledge, no longer subject to change. Using Fleck's own example, the etiological definition of syphilis as an infectious disease caused by the spirochete *T. pallidum* is unlikely to be abandoned, in contrast to Fleck's doubts which he expressed in so many words. Using arguments based on observations on unsymptomatic carrier status, and on the then uncertain phenotypic or genotypic stability of microbes as well as on the uncertainty of diagnostic procedures in bacteriology, Fleck suggests: "It therefore cannot be claimed that syphilis is definable epistemologically solely on the basis of *Spirochaeta pallida*. The idea of the syphilis agent leads to uncertainties attending the concept of bacteriological species as such and will thus depend upon whatever future developments there may be in this field". Clearly, as outlined above for Avery's demonstration of DNA as the hereditary principle (chapter 2), the concept of the bacteriological origin of syphilis has been retrospectively confirmed over and over again and passed whatever trial of strength it has been put to over the decades. The concept has become the foundation of antibiotic therapies which eventually became so effective as to abandon all threat formerly connected with the disease and, where they could be applied, to lead to its factual eradication. The etiologic thought-style may thus remain the final one in the history of syphilis, if only because it provided the solution to the problem. The case of syphilis may thus teach us a fifth epistemological category: The *communalization* of scientific knowledge. Once a problem seems to have been solved, it is incorporated into the body of human knowledge, but is dismissed as a scientific subject. The complex development of the fact is forgotten, it is perceived as a matter of course, obvious or even trivial.

Fleck's notion of thought-collective can be envisaged as an entity that shapes, albeit transiently, the thinking of an entire generation of scientists on a given theme of interest. More recently constructionist science critics focussed their interest on smaller units of scientific activity and noted that scientific facts are generated in "the laboratory", a special type of social environment with pronounced influence on ways in which facts are produced, interpreted, and disseminated in the scientific community. For Karin Knorr Cetina this field of epistemology, termed "laboratory studies", entails primarily two areas of relevance, scientific apparatus and politics[11,12]: "It allowed us to consider experimental activity within the wider context of equipment and symbolic practices within which the conduct of science is located.... Scientific objects are not only 'technically' manufactured in laboratories but are also inextricably symbolically or politically construed, for example through literary techniques of persuasion as one finds embodied in scientific papers, through the political stratagems of scientists in forming alliances and mobilizing resources, or through the selections and decision translations which 'build' scientific findings from within". Expanding on the distinction between observation and detection (see chapter 2), Knorr Cetina stresses the point that in the laboratory natural objects are not studied as they are. Rather, by subjecting them to scientific equipment natural objects are digitalized, fragmented, converted into images, purified versions, traces, electrical impulses, extractions, and the like. The type of equipment used for the study of a natural object thus reconfigures the object and the same object subjected to different equipment gets altered in different ways. Natural objects that may be transient, intact, insoluble, invisible, etc., are transformed to become persistent, fragmented, soluble, visible, etc. Particularly in the field of molecular biology, Knorr Cetina doubts that "the conclusions derived from such experiments are ... justified in terms of the equivalence of the experiment to real world processes". Like the objects of study, the scientists themselves become reconfigured in the laboratory. Unlike a computer, a scientist working in a laboratory gets transformed by persuasion and experience to think and act along certain avenues and to expect certain types of results when doing experiments. A circular relationship arises between procedure and outcome in that procedures get first appropriately chosen and then optimized in order to produce certain expected results. Thus the separation between experimenter and experiment is abolished, the idealistic doctrine of non-interference is violated, according to which the natural state of the object of experimentation and its course of events are not to be tampered with by the experimenter. Laboratories most successful in manipulation of objects and scientists acquire reputation and proliferate. Their leaders can spend much of their time representing, promoting, and recruiting junior personnel who are then trained to perpetuate the leader's attitude and career goal. Through the output of trained scientists, but also through the production of materials, reagents, specimen, etc., and their distribution to other groups, a laboratory may gain influence beyond its borders, its reconfigurations may get adopted by others and exert external dominance in an entire field of scientific activity.

A rather unique approach to the field of laboratory studies has been taken by the French philosopher Bruno Latour, published in 1979 under the title *Laboratory Life, The Construction of Scientific Facts* together with the British sociologist Steve Woolgar[13]. Essentially unfamiliar with biomedical research and laboratory work, Latour asked permission of Jonas Salk to spend an extended period of time in the Salk Institute in La Jolla. He obtained permission to observe scientists in their daily life and work in the department of Roger Guillemin, who shared the Nobel Prize in Medicine or Physiology for determining the structure of the hypothalamic hormone TRF (Thyrotropin Releasing Factor). Over a period of several months Latour first tried to understand the logistic aspects of people, apparatus and workspace in the lab. He described this phase of the study as being similar to the work of an anthropologist observing a primitive human tribe whose culture and customs initially appear strange and meaningless. According to Latour the position of a naive observer provides insights that could not be obtained by letting scientists themselves explain what they do, "...the observer would simply reiterate those accounts provided by scientists when they conduct guided tours of their laboratory to visitors".

After gradually coming to terms with what first seemed a disorderly and chaotic situation, Latour analyzed the techniques of information gathering and processing in the laboratory. His main conclusions from this phase of the study was that the laboratory workspace was divided into bench/equipment and desk areas, that the direction of information processing was from the bench to the desk and occurred by a cascade of what he dubbed "inscriptions": Technicians who had obtained material from animals prepared test tubes which they labeled, the labeled tubes went into a counter that printed a sheet of numbers, the numbers were fed into a computer which produced a curve, the curve went to the desk of a scientist who contemplated it by comparing it to other similar curves of previous experiments and to curves published by others in scientific journals. Finally, the curve was combined with other literary and graphic inscriptions and submitted for publication. While to an experimental scientist this may seem to some extent a trivial account, it should be noted that what Latour describes may be seen as the ultimate common denominator of the various forms of reconfigurations and transformations that natural objects undergo in the laboratory according to Knorr Cetina. Indeed, what we experimentalists study are the imprints on paper of the natural phenomena that we try to understand, often forgetful of the many transformations on the way.

Latour is in so far a realist philosopher as he concedes that science can generate knowledge about facts. In his analysis of the history of the elucidation of the structure of TRF, the existence of TRF is taken as unconditioned fact which, however, undergoes significant shifts in meaning as knowledge progresses in time. Over time, a fact such as TRF is processed by different "networks", in Latour's terminology not identical but equivalent to Fleck's thought-collective. Before 1962 TRF had been hypothetical, indicated by indirect evidence that the secretion of thyrotropin, a hormone produced by the anterior pituitary that stimulates the thyroid gland to produce thyroid hormone, is regulated by the brain. Between 1962 and 1966 efforts were made in a number of laboratories to isolate materials with TRF activity from the brain, resulting in preparations of increasing activity and purity. The structure was finally elucidated as a tripeptide (with blocked N-terminus and amidated C-terminus) in 1969 by two groups, Guillemain and Andrew F. Schally, Guillemain's former postdoc. The meaning of TRF thus shifted from a biological activity to an enriched fraction to a purified peptide to an amino acid sequence. Nowadays TRF is merely an established reagent used in endocrinological methodology.

Latour also carefully analyzed the publication and citation history uncovering the competition in the field, the strategies of establishing dominance, and the acquisition of credit. An important factor was the decision to go for the structure in the first place, as the conventional network of endocrinologists were either content with measuring biological activity of brain extracts in bioassays and/or considered it hopeless to try to prepare sufficient quantities of purified substance for structural studies, which would require the brains of thousands of rats. That left four laboratories in the race that had mastered to generate preparations from hypothalamus with high biological activity suggesting some enrichment of the active principle. The importance of the structural approach needed to be promoted in the scientific community and with granting agencies, flanked by the discreditation of any other approach that could have been taken. Two of the four laboratories, one from Japan and one from Chechoslovakia, dropped out of the race because of lack of resources, they could not raise the public funds necessary to upscale the preparation and equip the laboratory with the expensive machinery required for the work. In 1966 Schally had a preparation which had biological activity, consisting to 70% of three amino acids, which later turned out to be the ones of which TRF consists. The activity was protease sensitive, also consistent with a peptidic nature of the molecule. At the same time, on the basis of his own enzyme sensitivity studies, Guillemain had – erroneously – proclaimed that TRF was non-peptidic in nature. As a consequence, Schally did not have the confidence to proclaim his results as representing the amino acid composition of TRF and TRF as a peptide, which would have been correct and would have given him priority. From the numbers and contents of mutual citations Latour comes to the conclusion that Schally trusted Guillemain's results more than his own, whereas Guillemain was very much at ease with his own competence. Later, Guillemain obtained the same amino acid composition as Schally and the amino acid sequence of TRF was finally determined by synthesizing tripeptides in all possible permutations and terminal modifications, and testing them for biological activity. The results were published separately by both groups in 1969, and earned them both the Nobel Prize.

In his later work Latour had more of a problem than in *Laboratory Life* with recognizing sound scientific facts as what they are. In his "actor-network" model[14] he postulates that facts (actors) and networks (laboratories) validate one another, outside its network a fact becomes meaningless. Surprisingly, what philosophers do not seem to notice is the communalization of scientifically derived knowledge which proves beyond doubt that some scientific facts have left their networks and retained meaning in the real world. They seem to ignore the numerous instances of knowledge gained in the laboratory that has been subjected to and has passed the trial of strength in practical life. Constructivist philosophers either avoid this topic or summon rather bizarre arguments to deal with such cases. As an example of the successful exposure of a scientific fact to the challenges of the real world, Latour looked at the amazing success of Pasteur's anthrax vaccine in protecting farm animals and eradicating anthrax in rural France[15]. Latour explained this "miracle" by purporting that in the farm experiment in which the vaccine was first successfully examined Pasteur had managed "to transform the farm into a laboratory". While this might have been the case, Latour does not comment on the subsequent success of the vaccine upon application under field conditions. Ian Hacking, in his theory of self-vindication of laboratory sciences[16], also came across the problem that arises when the truth of facts generated in the laboratory is tested by real world application: "if mature laboratory sciences are self-vindicating, answering to the phenomena purified or created in the laboratory, how then are they generalizable?" Using – of all examples – the lack of precision of modern military aiming devices under war conditions, Hacking notes that "few things that work in the laboratory work very well in a thoroughly unmodified world – in a world that has not been bent towards the laboratory". Nevertheless, Hacking notes that anthrax and smallpox became eradicated in the real world (or parts thereof) owing to applied scientific knowledge. Here Hacking summons a "metaphysical mistake of thinking that truth or the world explains anything", and "it is not the truth or anything that explains the happy effects". What explains them then? It is obvious that neither Latour nor Hacking come to terms with the successful application of scientific facts in the real world.

Chapter 5
The science wars

Referees are volunteers, who as a whole put in a great deal of work for no credit, no money, and little or no recognition, for the good of the community. Sometimes a referee makes a mistake. Sometimes two referees make mistakes at the same time. ... I'm a little bit surprised that anyone is surprised at this. Surely you have seen bad papers published in good journals before this! ...referees give opinions; the real peer review begins after a paper is published.

Steve Carlip, physicist,
commenting on the Bogdanov affair

Humanists deal with literature, philosophy, law, etc., fields of knowledge that spring from the human mind. Natural scientists deal with gravitation, metabolism, earthquakes, microbes, etc., i.e. natural objects of which we have reason to think that they exist independent of the human mind, indeed independent of the existence of *Homo sapiens*. Needless to say, natural scientists have only their senses and brains to assess the objects of nature, so what results from their efforts are representations of natural objects formed in the human mind. However, unlike subjects of interest in the humanities, the natural objects under scrutiny of natural scientists are not, from scratch, products of the human mind. While until the previous century most philosophers of science had been trained as natural scientists, many contemporary science critics are humanists. They thus like to focus on products of the human mind and therefore concentrate, in their analyses of natural science, on the aspect of representation, often ignoring the independence of natural objects of the human mind. Natural scientists, in contrast, concentrate on natural objects, often forgetful of the transformations necessary in forming their representations. For natural science it would thus indeed be useful to learn more about the representational mechanisms in order to more precisely define and perhaps minimize their influence in understanding nature. For science scholars, conversely, it could be helpful to come to terms with their ignored and unsolved challenge that facts uncovered by natural sciences have indeed enabled mankind to reshape the real world, for better or worse but in an undisputable way. Instead, both sides have repeatedly engaged in fierce battles. The most recent example was Science Wars, which I will discuss here as it sheds light on the ways in which scientific disciplines deal with non-sensical or unreasonable productions in their own areas of professional expertise.

In 1959 the English chemist and science politician C.P. Snow, in his book *The Two Cultures*[1], observed a gulf between natural science and humanities "not only in an intellectual but also in an anthropological sense". He took the side of natural science, belittling humanists as a controversial and unproductive bunch of reactionaries who are natural enemies of the progress and welfare contributed by the natural sciences. Snow's book appeared at a time dominated by the competition between the US and the Soviet Union in space and military technology and the shock in the US about the Soviet launch of the first spacecraft, Sputnik. The ensuing international race for technical superiority led to a dominance of applied physics and technology not only in national propaganda and esteem but in public funding as well, a development that culminated in the "Star Wars" project of the 1980s. Conversely, events such as the Tchernobyl and Harrisburg catastrophes or the deleterious ecological effects of insecticides such as DDT raised concerns and opposition against the mentality that the future of the world was in good hands with the natural scientists, resulting in the sprouting of fundamentalistic and anti-scientific attitudes and a general polarization in the society about its relationship to science.

What is now referred to as the Science Wars started with the book *Higher Superstition, The Academic Left and Its Quarrels With Science*, by the biologist Paul R. Gross and the mathematician Norman Levitt, which appeared in 1994[2]. The book was an open attack on the modern science critics, focussing on the proponents of relativism and constructivism and accusing them of ignorance about the scientific theories they discussed. Predictably, these points were vehemently rebutted by scholars of science studies, accusing the authors of failure to understand the theoretical approaches they criticized[3]. Together with Gerald Holton, Gross and Levitt organized a conference, entitled "The Flight from Science and Reason", hosted by the New York Academy of Sciences in 1995, in which the majority of participants expressed a critical attitude towards the ways in which humanist intellectuals dealt with natural sciences[4].

The style of *Higher Superstition* was that of a pamphlet, attacking not only *bona fide* science philosophers but including what the authors summarized derogatively under "academic left", i.e. feminism, afrocentrism, fundamental ecologism, alternative medicine, and the like. This brought into sharp focus the arising public hostility towards science as being fed by the distorted arguments that fundamentalists derived from the works of science critics, for example those advertised by the "intelligent design" creationist movement to discredit Darwinian evolution. Gross and Levitt's book was flanked by several critical contributions accusing feminism of deterring women from proper learning and research and of disguising opinions as facts[5]. Along similar lines, afrocentrism was accused of making false claims about its role in the origin of European culture[6,7].

Science Wars reached its peak when the physicist Alan Sokal submitted a paper entitled: "Transgression of the boundaries: Toward a transformative hermeneutics of quantum gravity"[8] to *Social Text*, a prominent US journal of cultural critique and social science. The editors of the journal were collecting articles to compose a special issue on Science Wars as a forum for the science critics to answer to the assaults of *Higher Superstition* and its ilk. Sokal's article had not been invited but was accepted for publication without peer review, although Sokal was the only natural scientist among the authors, and not known to have published philosophical texts. Simultaneously with the appearance of the Science Wars issue of *Social Text*, Sokal disclosed in another periodical, *Lingua Franca*, that his article in *Social Text* was a hoax, a non-sensical text deliberately stuffed with phrases and jargon in current use among science critics[9]. Sokal's aim was to test weather "a leading North American journal of cultural studies – whose editorial board includes luminaries like Fredric Jameson and Andrew Ross – (would) publish an article liberally salted with nonsense if (a) it sounded good and (b) it flattered the editor's ideological preconceptions?" As the answer was yes, the "more or less explicit rejection of the rationalist tradition of enlightenment" by "postmodern" science critique was thus exposed as uncritical, superficial, lacking seriousness and credibility.

The editors of *Social Text* defended their decision and accused Sokal of making ridicule, as another form of censorship, of their serious scientific effort. The affair caused quite a turmoil in the intellectual community, with prominent individuals standing up for one or the other side[10]. For example physicist Steven Weinberg complimented Sokal for directing public attention to a conflict of imminent urgency and demanded a strict separation between rational scientific reasoning and the irrational tendencies of science studies. Conversely, social scientist Joan Fujimura sided with *Social Text* and used the historical challenge of Euklidian geometry by Carl Friedrich Gauss as an example for the suppression of a scientifically justified criticism[11]. The affair reached even the international mainstream press, with comments mostly critical of the performance of the *Social Text* editors. The London Times commented that *Social Text* was not a scientific but "…a political magazine…: under appropriate circumstances it is prepared to let agreement with its ideological orientation trump over other criterion for publication, including something as basic as sheer intelligibility".

With the Sokal affair natural science had a welcome argument in dealing with the attack of science studies. However, not much later a similar affair, this time concerning theoretical physics, shed an equally dim light on

The science wars
41

the ability of natural scientists to recognize pseudo-scientific nonsense if it was camouflaged as serious contribution. The Bogdanov affair (ref.[12–15]) is in part the reverse of the Sokal affair, and refers to a scientific dispute about a series of publications in international physics journals dealing with the events in the early phase of the big bang, known as the Planck era of the origin of the universe. The Bogdanov brothers, twins born in 1949 of immigrant parents in France, had studied mathematics in Paris. In the 1980s they became well known in France as producers of a popularized science show called TempsX on French television. In 1991 they published a best-selling science book entitled *Dieu et la Science* on the cover of which they presented themselves with doctoral titles, which neither of them in fact held. This was uncovered by a journalist, as well as the fact that much of the book had been plagiarized from an earlier book by astronomy professor Trin X. Thuan of the University of Virginia. The ensuing accusations forced the brothers, well into their 40s, to enlist as doctoral students at the University of Burgundy in order to work for PhD degrees. In 1999 one brother, Grichka, passed the defense of his thesis: "Quantum fluctuations of the signature of the metric at the Planck scale", though with a low passing grade. The other, Igor, failed but was given the chance to prepare a new thesis, entitled "Topological state of spacetime at the Planck scale". His supervisor accepted the thesis on the condition that Igor could publish his results in three peer-reviewed articles. This accomplished, he obtained the degree in 2002, also with low grade. The brothers developed their theory on the origin of the universe in a total of five papers (ref.[16–20]) which appeared in peer-reviewed physics journals including the respectable *Annals of Physics* and *Classical and Quantum Gravity*. Some of the papers are virtually identical to each other.

The first to notice the pseudoscientific nature of the Bogdanov papers was the German physicist Max Niedermaier, then working in the University of Tour, France, who suspected that these papers were deliberate nonsense of the Bogdanovs, attempting to reveal weaknesses in the peer review system of physics journals. Niedermaier alerted some fellow physicists by an e-mail which subsequently passed around the world and was the starting point of the scandal. Soon there was general agreement among theoretical physicists that the Bogdanovs' papers were non-sensical and meaningless[21]. A few quotes from a Usenet newsgroup established to discuss the matter: "a mishmash of superficially plausible sentences containing the right buzzwords in approximately the right order. There is no logic or cohesion in what they write."; "the papers consist of buzzwords from various fields of mathematical physics, string theory and quantum gravity, strung together into syntactically correct, but semantically meaningless prose"; and so on.

Unlike Sokal, the Bogdanovs rejected any allegation of deliberately publishing nonsense and insisted that they had succeeded in unravelling the most important questions about the beginning of space-time. Their work, however, was not understood by other physicists because of its complexity and specialization. Implying that the Bogdanovs were indeed in good faith about their own research, physicist Peter Woit, author of the book "Not Even Wrong", wrote[22]: "The Bogdanovs' work is significantly more incoherent than just about anything else being published. But the increasingly low standard of coherence in the whole field is what allowed them to think they were doing something sensible and to get it published".

The ensuing discussion soon swept over the boundaries of the academic scene and received broad coverage in the public media. As it turned out, both the theses and the papers had passed the peer review system without problems. For example, the reviewers of the theses had not been overly enthusiastic but attested that the work was "of interest" and contained "stimulating new ideas". Similarly, reviewers found one paper "sound, original, and of interest", another paper "interesting as a possible approach of the Planck scale physics". Most referees requested revisions, indicating that the reviewing was done with some degree of attention and accuracy. Revisions were done as requested and the papers accepted. Only in one case had a reviewer recommended rejection, which was not followed by the editors.

For the question of validity of scientific knowledge it is of interest to ask if the Bogdanov papers had any damaging effects, or any impact at all on physics. It certainly had in the public domain, as certain mundane newspapers published articles that tried to ridicule modern physics as a whole. In science itself, however, the impact was negligible. Up to 2007 the databanks mention a total of six citations for the Bogdanovs' publications. Four of them are citations among themselves and only two are by other physicists. The impact of the work in physics was just close to nil, a comforting ascertainment as far as the process of selection by retrospective confirmation is concerned. However, the majority of sound scientific papers receives low quotation numbers as well, between 0 and 20 is normal, only a few make it close to 100 and even less to several hundred citations. The impact of scientific contributions, as judged by citation rates, while it does not distinguish between wrong and boring, clearly identifies some productions as important and others as inconsequential.

References and further reading, Part I

Chapter 2

References

1. Laertius D (1970) *Lives of Eminent Philosophers*. Loeb Classical Library, Cambridge, Mass
2. Wagner G (2001) *Auguste Compte zur Einführung*. Junius, Hamburg
3. Fuchs-Heinritz W (1998) *Auguste Compte. Einführung in Leben und Werk*. Westdeutscher Verlag, Opladen/Wiesbaden
4. Kolakowski L (1972) *Positivist Philosophy – From Hume to the Vienna Circle*. Penguin Books, London, UK
5. Kant I (1781) *Die Kritik der reinen Vernunft. Prolegomena. Drittes Hauptstück. Metaphysische Anfangsgründe der Mechanik*. Available at: www.ikp.uni-bonn.de/kant/aa04/
6. Reichenbach H (1935) *Wahrscheinlichkeitslehre*. Leiden, Holland
7. Kraft V (1953) *The Vienna Circle: The Origin of Neo-positivism, a Chapter in the History of Recent Philosophy*. Greenwood Press, New York
8. Schlick M (1925) *Allgemeine Erkenntnislehre*. Verlag von Julius Springer, Berlin
9. Stove DC (1973) *Probability and Hume's Inductive Scepticism*. Oxford University Press
10. Miller DS (1949) Hume's Deathblow to Deductivism. *The Journal of Philosophy* Vol. XLVI, No. 23
11. Hume D (1893) *An Enquiry Concerning Human Understanding*, 1748, ed. Selby-Bigge, Oxford University Press, Oxford
12. Hume D (1739) *A Treatise of Human Nature*. ed. Selby-Bigge, Oxford University Press, 1888, Oxford
13. Pickering A (ed) (1992) *Science as Culture and Practice*. The University of Chicago Press, Chicago and London; Articles by: Pickering A and Stephanides A, Knorr Cetina K, Hacking I, Fujimura J, Woolgar S
14. Quine WV (1953) *From a Logical Point of View*. Harper Torchbooks, New York
15. Duhem PM (1954) *The Aim and Structure of Physical Theory*. Princeton University Press
16. Gardner M (1983) *The whys of a philosophical scrivener*. Quill, New York
17. Feyerabend PK (1975) *Against Method*. NLB, London
18. Bonk T (2008) *Underdetermination. An essay on evidence and the limits of knowledge*. Springer, Netherland
19. Hesse M (1980) *Revolutions and Reconstructions in the Philosophy of Science*. Indiana University Press
20. Bloor D (1991) *Knowledge and Social Imagery*. Routledge, Chicago University Press
21. Feyerabend PK (1970) Consolations for the Specialist. In: Lakatos, Musgrave (eds): *Criticism and the Growth of Knowledge*. Cambridge University Press
22. Feyerabend PK (1978) *Science in a Free Society*. NLB, London
23. Hübner K (1993) *Kritik der Wissenschaftlichen Vernunft*. Verlag Karl Alber, Freiburg

44 Chapter 5

24 Hübner K (1985) *Die Wahrheit des Mythos*. Beck, München
25 Avery OT, MacLeod C, McCarty M (1944) Studies on the chemical nature of the substance inducing transformation of pneumococcal types. *J Exp Med* 79: 137–158
26 Griffith F (1928) The significance of pneumococcal types. *J Hyg* 27: 113
27 Hershey AD, Chase M (1952) Independent functions of viral protein and nucleic acid in growth of bacteriophage. *J Gen Physiol* 36(1): 39–56
28 Watson JD, Crick FH (1953) Molecular structure of nucleic acids; a structure for deoxyribose nucleic acid. *Nature* 171 (4356): 737–738
29 Meselson M, Stahl FW (1958) The replication of DNA in *Escherichia coli*. *Proc Natl Acad Sci USA* 44: 671–682
30 Nirenberg MW, Matthaei JH (1961) The dependence of cell-free protein synthesis in *E. coli* upon naturally occurring or synthetic polyribonucleotides. *Proc Natl Acad Sci USA* 47: 1588–1602
31 Crick FH, Barnett L, Brenner S, Watts-Tobin RJ (1961) General nature of the genetic code for proteins. *Nature* 192: 1227–1232
32 Miescher JF (1871) Ueber die chemische Zusammensetzung der Eiterzellen. *Medisch-chemische Untersuchungen* 4: 441–460

Further reading

McCarty M (1985) *The transforming principle – discovery that genes are made of DNA*. W.W. Norton Comp, New York
Descarte R (1966) *Philosophical Writings*, ed. Anscombe GEM and Geach P, Nelson, London
Bloor D (1998) The Strengths of the Strong Programme. *Philosophy of the Social Sciences* 11: 173
Carnap R (1950) *Logical Foundations of Probability*. University of Chicago Press
Hume D (1882) Essays, Moral, Political and Literary, 1742, in *David Hume, The Philosophical Works*, ed. Green H, Grose HH, google.books, London
MacIntyre A (1969) Hume on 'Is' and 'Ought'. In: Hudson (ed): *The Is-Ought Question*. MacMillan, London
Stove DC (1965) Hume, Probability, and Induction. *The Philosophical Review* Vol. LXXIV, No. 2
Stove DC (1975) Hume, the Causal Principle, and Kemp Smith. *Hume Studies* Vol. I, No. 1

Chapter 3

References

1 Bronowski J (1974) Humanism and the Growth of Knowledge. In: Schilpp (ed): *The Philosophy of Karl Popper*. Open Court, La Salle
2 Schilpp PA (ed) (1974) *The Philosophy of Karl Popper*. LaSalle, Illinois, USA
3 Popper KR (1959) *The Logic of Scientific Discovery*. Hutchinson, London
4 Einstein A (1905) Zur Elektrodynamik bewegter Körper. *Annalen der Physik* 17: 891–921

5 Einstein A (1916) Die Grundlage der allgemeinen Relativitätstheorie. *Annalen der Physik* 49: 769–822
6 Crick FHC (1958) On Protein Synthesis. *Symp Soc Exp Biol* XII: 139–163
7 Crick FHC (1958) Central Dogma of Molecular Biology. *Nature* 227: 561–563
8 Baltimore D (1971) Reversal of information flow in the growth of RNA tumor viruses. *N Engl J Med* 284(5): 273–275
9 Coffin JM, Temin HM (1971) Ribonuclease-sensitive deoxyribonucleic acid polymerase activity in uninfected rat cells and rat cells infected with Rous sarcoma virus. *J Virol* 8(5): 630–642
10 Prusiner SB (1982) Novel proteinaceous infectious particles cause scrapie. *Science* 216: 136–144
11 Crick F (1990) *What Mad Pursuit: A Personal View of Scientific Discovery.* Basic Books reprint edition, New York
12 Silverstein AM (1989) *A history of immunology.* Academic Press, Inc. San Diego, New York
13 Kindt TJ, Capra JD (1984) *The antibody enigma.* Plenum Press, New York
14 Foote J, Milstein C (1994) Conformational isomerism and the diversity of antibodies. *Proc Natl Acad Sci USA* 91: 10370–10374
15 Söderquist T (2003) *Science as autobiography, the troubled life of Niels Jerne.* Yale University Press, New Haven, London
16 Lakatos I (1970) Falsification and the Methodology of Scientific Research Programmes. In: Lakatos, Musgrave (eds): *Criticism and the Growth of Knowledge.* Cambridge University Press
17 Lakatos I (1976) *Proofs and Refutations.* Cambridge University Press
18 Kuhn TS (1970) *The Structure of Scientific Revolutions.* University of Chicago Press
19 Kuhn TS (1970) Logic of Discovery or Psychology of Research. In: Lakatos, Musgrave (eds): *Criticism and the Growth of Knowledge.* Cambridge University Press
20 Barber B (1961) Resistance by Scientists to Scientific Discovery. *Science* 134: 569
21 Bruner JS, Postman L (1949) On the Perception of Incongruity: A Paradigm. *Journal of Personality* 18: 206
22 Hacking I (1992) The Self-Vindication of Labotatory Sciences. In: Pickering A (ed): *Science as Culture and Practice.* The University of Chicago Press, Chicago and London
23 Laudan L (1996) *Beyond Positivism and Relativism. Theory, Method, and Evidence.* Westview Press, Oxford, UK

Further reading

Popper KR (1963) *Conjectures and Refutations.* Routledge and Kegan Paul, London
Popper KR (1968) Theories, Experience, and Probabilistic Intuitions. In: Lakatos (ed): *The Problem of Inductive Logic.* North-Holland, Amsterdam
Popper KR (1970) Normal Science and its Dangers. In: Lakatos, Musgrave (eds): *Criticism and the Growth of Knowledge.* Cambridge University Press
Popper KR (1972) *Objective Knowledge.* Oxford University Press
Popper KR (1974) Autobiography. In: Schilpp (ed): *The Philosophy of Karl Popper.* Open Court, La Salle

Stove DC (1978) Popper on Scientific Statements. *Philosophy* Vol. 53: 81–88
Popper KR (1957) Philosophy of Science: a Personal Report. In: Mac (ed): *British Philosophy in the Mid-Century*. Allen and Unwin, London
Kuhn TS (1977) *The Essential Tension*. University of Chicago Press
Kuhn TS (1957) *The Copernican Revolution: Planetary Astronomy in The Development of Western Thought*. Cambridge University Press, Cambridge, Mass
Kuhn TS (1959) Conservation of Energy as an Example of Simultaneous Discovery. In: Clagett M (ed): *Critical Problems in the History of Science*. Madison, Wis
Kuhn TS (1961) The Function of Measurement in Modern Physical Science. *Isis* 52: 162–193

Chapter 4

References

1   Fleck L (1935) *Entstehung und Entwicklung einer wissensachaftlichen Tatsache; Einführung in die Lehre vom Denkstil und Denkkollektiv*. Benno Schwabe und Co. Verlagsbuchhandlung, Basel
2   Trenn TJ, Merton RK (eds) (1979) *Ludwik Fleck – Genesis and Development of a Scientific Fact*. University of Chicago Press, Chicago/London
3   Reichenbach H (1938) *Experience and Prediction. An Analysis of the Foundations and the Structure of Knowledge*. University of Chicago Press, Chicago
4   Kuhn T (1996) *The Structure of Scientific Revolutions*. University of Chicago Press, Chicago
5   Cohen RS, Schnelle T, Reidel D (1986) *Cognition and fact – Materials on Ludwik Fleck*. Publishing Co. Dordrecht
6   Szaniawski K (ed) (1988) *The Vienna Circle and the Lvov–Warsaw School*. Kluwer, Dordrecht/Boston/London
7   Some specific properties of the medical style of thinking (orig. in Polish), *Archivum Historji i Filosofji oraz Historji Nauk Pozyronigczych* 6, 1927; Zur Krise der Wirklichkeit. *Naturwiss* 17, 1929, 425; Specific observation and perception in general (orig. in Polish), *Przegladf Filozoficzny* 38, 1935; The problem of epistemology (orig. in Polish), *Przegladf Filozoficzny*, 39, 1936. English translations in ref. 5
8   Kogon E (1946) *Der SS Staat, Das System der Deutschen Konzentrationslager*. Verlag der Frankfurter Hefte, Munich
9   Fleck L (1946) Problems of the science of science. *Zykie Nauki 1* (orig. in Polish, English translation in ref. 5)
10 Fleck L (1947) *To look, to see, to know. Problemy* (orig. in Polish, translation in ref. 5)
11 Knorr Cetina K (1992) The couch, the cathedral, and the laboratory. On the relationship between experiment and laboratory in science. In: Pickering A (ed): *Science in practice and culture*. University of Chicago Press
12 Knorr Cetina K (1981) *The Manufacture of Knowledge, An Essay on the Constructivist and Contextual Nature of Science*. Pergamon Press, Oxford

References and further reading, Part I

13 Latour B, Woolgar S (1986) *Laboratory life: The construction of scientific facts.* 2nd ed. Princeton University Press
14 Latour B (2005) *Reassembling the Social: An Introduction to Actor-Network-Theory.* Oxford University Press, Oxford
15 Latour B (1988) *The Pasteurization of France.* Cambridge University Press, Cambridge, Mass
16 Hacking I (1992) The Self-Vindication of Labotatory Sciences. In: Pickering A (ed): *Science as Culture and Practice.* The University of Chicago Press, Chicago and London

Further reading

Knorr Cetina K, Mulkay M (eds) (1983) *Science Observed: Perspectives on the Social Study of Science.* Sage, London
Latour B (1996) On Actor Network Theory. A Few Clarifications. *Soziale Welt* 47: 369–381

Chapter 5

References

1 Snow CP (1993) *The Two Cultures.* Cambridge University Press
2 Gross P, Levitt P (1994) *Higher Superstition: The Academic Left and its Quarrels with Science.* Johns Hopkins University Press, Baltimore
3 Flower MJ (1995) Review of Higher Superstition. *Contemporary Sociology* Vol. 24, No. 1: 113–114; see also *Isis* Vol. 87, No. 2 (1996); *American Anthropologist* Vol. 98, No. 2 (1996); *Social Studies of Science* Vol. 26, No. 1 (1996); *The Journal of Higher Education* Vol. 66, No. 5 (1995)
4 Gross PR, Levitt N, Lewis MW (1997) *The Flight from Science and Reason.* New York Academy of Science, New York
5 Patai D, Koertge N (1994) *Professing Feminism: Cautionary Tales from the Strange World of Women's Studies.* Basic Books, New York
6 Lefkowitz M (1996) *Not Out of Africa: How Afrocentrism Became an Excuse to Teach Myth As History.* HarperCollins, New York
7 Grant M (1995) *Greek and Roman Historians: Information and Misinformation.* Routledge, Florence, Kentucky, USA
8 Sokal A (1996) Transgressing the Boundaries: Toward a Transformative Hermeneutics of Quantum Gravity. *Social Text* 46/47, Vol. 14, Nos. 1 & 2: 217–252
9 Sokal A (1996) A Physicist Experiments with Cultural Studies. *Lingua Franca* May/June: 62–64
10 Ashman KM, Barringer PS (eds) (2001) *After the science wars.* Routledge, London, UK
11 Fujimura J (1998) Authorizing Knowledge in Science & Anthropology: Comparison with 19th Century Debate on Euclid. *American Anthropologist* 100, No. 2: 347–360
12 Monastersky R (2002) French TV Stars Rock the World of Theoretical Physics. *The Chronicles of Higher Education,* Nov. 2

13 Overbye D (2002) Are They a) Geniuses or b) Jokers? French Physicists Cosmic Theory Creates Big Bang on Its Own. *The New York Times,* Nov. 9
14 Butler D (2002) Theses spark twin dilemma. *Nature* 420 (6911): 5
15 Sokal A, Bricmont J (2003) *Intellectual Impostures*. Profile Books, London
16 Bogdanov G, Bogdanov I (2001) Topological field theory of the initial singularity of spacetime. *Classical and Quantum Gravity* 18 (21): 4341–4372
17 Bogdanov I (2001) Topological origin of inertia. *Chechoslovak Journal of Physics* 51: 1153–1176
18 Bogdanoff G, Bogdanoff I (2003) Thermal equilibrium and KMS condition at the Planck scale. *Chinese Annals of Mathematics Series B* 24: 267–274
19 Bogdanoff G, Bogdanoff I (2002) KMS space-time at the Planck scale. *Nuove Cimenta della Societa italiana di Fisica B* 117: 417–424
20 Bogdanoff G, Bogdanoff I (2002) Spacetime metric and the KMS condition at the Planck scale. *Annals of Physics* 296: 90–97
21 see usenet google groups: sci.physics.research
22 Woit P (2006) *Not Even Wrong: The Failure of String Theory and the Search for Unity in Physical Law*. Basic Books, New York

Further reading

Collins H, Pinch T (1993) *The Golem: What every one should know about science.* Cambridge University Press, Cambridge

Nelkin D (1996) The Science Wars: Responses to a Marriage Failed. *Social Text* 46/47, Vol. 14, Nos. 1 & 2: 93–100

Parsons K (ed) (2003) *The Science Wars: Debating Scientific Knowledge and Technology*. Prometheus Books, Amherst

Baringer PS (2001) Introduction: 'the science wars'. In: Ashman KM, Baringer PS (eds): *After the Science Wars.* Routlege, New York

Part II
Origins, Rise, and Fall of the Network Paradigm

Chapter 6

The immune system, pre-network paradigms

Paradigms gain their status because they are more successful than their competitors in solving a few problems that the group of practitioners has come to recognize as acute. To be more successful is not, however, to be completely successful.... The success of a paradigm is at the start largely a promise of success...
T. S. Kuhn,
The Structure of Scientific Revolutions

Species of living creatures can only be successful in evolution, i.e. multiply to large numbers and exist for extended periods of time, if they are adapted to a habitat, i.e. are able to make use of special resources in the habitat for shelter and nourishment. Other species, if they compete for the same resources in the same habitat, but do so less successfully, are eliminated by starvation and slower reproduction, if not by direct killing. A special type of competition involves the use of one species by another as its habitat. For example, multicellular organisms, from plants to mammals, provide a very good habitat for small organisms such as bacteria, parasites, or fungi, which are adapted to feed on the substances of which multicellular organisms largely consist, i.e. proteins, polysaccharides, and lipids. Viruses, without life by themselves, only begin to live after they enter living cells of a suitable host. Some of these microorganisms make use of the resources provided by their multicellular hosts without harming them, and are thus harmless commensals. Others even provide essential metabolites that the host organism is unable to produce and are thus useful symbionts. A large number of microorganisms, however, cause diseases and some can kill a multicellular host, they are pathogens. There are many intermediate relationships, for example certain microbes are harmless for a healthy host but can harm a host that is compromised in its health, so-called facultative pathogens. The evolutionary success of multicellular organisms thus depends on their ability to deal with the constant onslaught of a multitude of parasitic microbes[1].

Multicellular organisms have therefore been selected in evolution by their ability to live and reproduce in the presence of a large variety of potentially dangerous microorganisms. To this end, they have developed physiological defense systems which keep the microorganisms from overwhelming them. For example, plants possess a mechanism, termed hypersensitive cell death, by which all cells in the vicinity of a microbial invasion rapidly die so that the invading microorganisms do not find living cells to feed on and or multiply within. In the animal kingdom a number of different defense molecules have been identified. For example, insects produce certain peptides that are toxic for certain bacteria. Fungi produce the well known antibiotics many of which competitively inhibit certain essential metabolic steps in bacteria and thus keep them from multiplying. What makes this a continuing struggle is the fact that microorganisms usually have very short generation times and thus undergo rapid changes by mutation and selection, continuously generating new variants that can circumvent the defense systems of their host. Lower animal species that rely on defense systems consisting of a limited number of molecules or mechanisms must have a hard time adapting to the changing world of microbes, and it is so far not clear how they manage. Certainly, higher organisms such as vertebrates and mammals, because of their much longer generation times, are unable to counteract genetic variation of the microorganisms by generating their own mutations. So from a certain size and level of biological complexity the static defense systems used by most invertebrates became insufficient for effective protection. Thus, with the advent of the vertebrates a novel defense system evolved which, unlike the earlier ones, was able to anticipate any type of present and future pathogenic invader. Anticipation was achieved by novel genetic mechanisms that generate a sheer endless variety of receptor proteins, a strategic repertoire so large that any possible molecular shape in nature is covered with a high probability. These receptors arm an array of defensive cell types so as to enable them to identify, capture, and deal with, foreign invaders of any possible structure.

The idiotypic network theory was the central theory and became the leading paradigm in immunological research from about 1975 to 1990. In order for the reader to appreciate its historical role in studies of the immune system, I must first outline some concepts on the molecular and cellular mechanisms of immunity in pre-network times. Immunologists of my generation may be best advised not to read this and the next chapter, and younger colleagues interested in the history of their science better read the far more comprehensive treatment of Arthur Silverstein[2]. This minimal historical sketch of immunological paradigms only includes those that appear to me as absolutely essential for non-immunologists to understand the derivation of the network theory. It is not the purpose of this book to present a comprehensive history of immunology and, anyway, how did

Ludwik Fleck put it? "It is difficult, if not impossible, to give an accurate historical account of a scientific discipline"[3].

From the end of the 19th century until its solution about 1975, the antibody problem[4] was in the center of interest of most immunologists. The phenomenon of acquired resistance to infectious agents had been known for centuries in many parts of the world and had led to prophylactic vaccination procedures for diseases such as smallpox very early on. Towards the end of the 19th century investigators such as Louis Pasteur, Robert Koch, Theobald Smith, and others had established the germ theory of disease and Pasteur in particular used this knowledge for the production of effective vaccines by attenuation of infectious agents[2]. However, these investigators took the ability of the mammalian host to acquire resistance for granted and did not attempt to study the host's defense mechanisms. Their interest was fully occupied by the microbes they tried to manipulate, they were medical microbiologists rather than immunologists. It was thus the discovery of antibodies which I see as the starting point of immunology as a mature science.

Evidence for the existence of antibodies had been first reported in 1890 by the German Emil von Behring together with his Japanese coworker Shibasaburo Kitasato[5], and their observations gave rise to the first unifying paradigm in immunology, namely that specific immunity is associated with the formation of antibodies*. They had noted that animals either infected with pathogenic bacteria or injected with bacterial toxins acquired an activity in their blood serum that, when transferred into non-infected animals, could protect them from infection with the bacteria producing the active toxin. They were the first to note the phenomenon of specificity, i.e. immune sera to diphtheria toxin neutralized diphtheria toxin but not tetanus toxin, while immune sera to tetanus toxin neutralized tetanus toxin but not diphtheria toxin. They also demonstrated the phenomenon of induction, i.e. the appearance of anti-toxin activity in the serum required immunization with the appropriate toxin. The knowledge that von Behring and Kitasato developed in their experiments was in due course applied in practical medicine. Larger animals including horses were immunized with the medically relevant bacterial toxins and their sera served to rescue, among others, countless children infected with diphtheria, as well as thousands of World War I soldiers infected with tetanus. In the pre-antibiotics era these were the first and only causative therapies for several devastating human infections, earning von Behring in 1901 the first Nobel Prize in Medicine or Physiology.

---

* To say that von Behring and Kitasato discovered antibodies is a gross oversimplification. Rather, the generation of the antibody paradigm is an interesting example for the thesis of this book, namely that scientific knowledge emerges from a mass of initially fictional notions, of which only a minority turns into fact. This occurs by a lengthy multistep process which will be discussed in Part III.

54   Chapter 6

In the following decades similar inducible activities in immune sera were discovered which, depending on the inducing agent and the procedures used for detection, were given different names. For example, those that were induced by immunization with red blood cells and detected by agglutination reactions were called agglutinins, those that formed a precipitation reaction with a soluble antigen were called precipitins, and so on. The term "antibody" eventually emerged, collectively describing all of these activities, as well as the term "antigen" for the inducing substances*. Eventually it was realized that antibodies could be induced to any type of natural antigen, soluble or particular, living or non-living, infectious or harmless, no matter what. The presence of antibodies led to a state of immunity, which was particularly impressive in the case of infections to which immunity could be acquired either by prophylactic immunization, termed active immunity, or by transfer of immune serum containing the appropriate antibodies, termed passive immunity.

The first comprehensive theory of antibody formation was put forward in 1900 by Paul Ehrlich[6]. It became known as the "side chain theory" and earned Ehrlich the 1908 Nobel Prize in Physiology or Medicine. Ehrlich postulated that all cells in the body, not only lymphoid cells, carried "atomic groups" on their surface which he termed side chains. The side chains had different molecular shapes, enabling them to specifically react with foreign molecules of various chemical nature. Important here was the concept of specific interactions between different molecular entities, which Ehrlich had deduced from his studies on the specific interactions of dyes with tissues or cells. He had worked with a panel of dyes, such as methylene blue, which not only specifically stained tissues or cellular structures but also were used as therapeutics in infections such as malaria. Ehrlich suspected the origin of specificity of the dyes to lie in their ability to engage in weak interactions when a part of their molecular surface happens to interact with part of the target molecule so that both surfaces fit onto one another. In the case of dyes, this fit allows the formation of salt bridges which result in a feeble binding, whereas the stronger binding of antibodies suggested the formation of a chemical reaction. Later Ehrlich applied similar concepts to chemotherapy and developed salvarsan for the treatment of syphilis[7]. Non-covalent binding between different biomolecules on the basis of molecular complementarity is now known to be one of the most fundamental principles on which biological functions and structures are based. Before the concept of complementarity was developed, Ehrlich must have had similar rules in mind for the fit between the side chains and their ligands[8].

Soluble antibodies are induced by excessive ligand binding to the side chains, causing the cell to synthesize more of the same side chain until the surface density gets too high and the side chains are shed into the blood.

---

*   The erratic emergence of the terms "antigen" and "antibody" will be discussed in Part III.

Ehrlich's side chain theory thus postulated that all antibodies are preformed in the body as cell surface receptors. So the question arose how many different antibodies the vertebrate organism can produce. This question became particularly pressing when it was found that most antibodies represent just one type of protein, γ-globulins[9]. Proteins consist of chains of amino acids, so-called polypeptide chains. These are strings of amino acids, of which there are 20 different ones in nature, which are connected by peptide bonds that link the carboxyl group of one amino acid to the amino group of the adjacent one. Polypeptide chains are thus asymmetric with a free amino group at the so-called amino terminus and a free carboxyl group at the carboxy terminus. Polypeptide chains may vary in length, amino acid sequence, and molecular folding, and a protein may consist of one or more polypeptide chains, so that an endless variety of different proteins is possible. A critical experiment was contributed by Karl Landsteiner, recipient of the 1930 Nobel Prize for the discovery of human blood groups, who showed that antibodies can be produced by immunization with chemically synthesized artificial molecules that do not even occur in nature[10]. This indicated that the number of possible antibodies must be larger than anticipated, i.e. huge, putting into question that all of them could be preformed in the body. The knowledge of genetics was very fragmentary at the time, so that the relationship between genes and antibodies was open to speculation and controversy. Based on the reasoning that the number of proteins that an organism can synthesize must be limited, so-called "instructional" theories[11,12] were put forward which postulated that the antibody polypeptide chains were initially flexible and would fold around the antigen upon biosynthesis or on first contact, so as to acquire a complementary shape which was then maintained, i.e. antibody specificity was instructed by the antigen. The instructional theories postulated that one antibody polypeptide chain could be folded into many different conformations, and thus required only one or a small number of preformed antibody polypeptide chains. The instructional theory was the prevailing paradigm on antibody specificity from about 1930 to 1955.

During this interval knowledge was rapidly acquired in several biological fields, supportive of but also arguing against the instructional generation of antibody specificity. In 1944 Oswald Avery and coworkers reported[13] to have identified DNA as the transforming principle of pneumococci and subsequent experiments retrospectively augmented these results to mean that DNA was indeed the universal substance of heredity (see chapter 2). In 1953 Watson and Crick published their model of the three-dimensional structure of DNA[14], and suggested plausible mechanisms of how the DNA molecule would replicate and code for proteins. Accordingly, chromosomes were viewed to consist of long threads of DNA containing the genes like beads of a necklace. The notion: one gene – one polypeptide arose as the central principle of the relationship between genes and proteins. Estimates of how many genes the vertebrate organism would contain were nevertheless difficult to make but indirect arguments pointed at numbers in the order of 10^5, certainly less than 10^6. Supporters of the instructional theory could use these estimates to claim that there were not enough genes to encode all possible antibody polypeptide chains.

Progress was also made in protein chemistry and antibody structure. A compelling result was that antibodies whose conformation and antigen binding specificity was destroyed by chemical denaturation regained their original specificity upon renaturation. With the introduction of X-ray crystallography to protein chemistry the three-dimensional structures of proteins could be elucidated. This was initially restricted to only a few biomolecules which could be obtained both in large quantities and in very pure state, which was not the case for antibodies. Nevertheless, the initial results suggested that proteins such as myoglobin have a very defined conformation, which is the same for all molecules in a crystal. For most proteins it seemed that the primary amino acid sequence unequivocally determined the secondary and tertiary structures, while others such as enzymes may have two but not multiple conformations. Multiple possible conformations of one polypeptide as postulated for antibodies by the instructional theory became thus to some extent doubtful.

Upon electrophoretic analysis of serum proteins antibodies were found in the gamma-globulin fraction and were thus also called immunoglobulins or gamma-globulins[9]. Gerald Edelman at the Rockefeller Institute showed that gamma-globulins consist of two types of polypeptide chains, termed heavy and light chains because the former is about twice as long as the latter[15]. Rodney Porter in Britain had shown that upon digestion with proteases antibodies fall into three fragments of about equal size, two of which maintain the ability to attach to antigens[16]. These results, for which both shared the Nobel Prize in 1972, led to the well-known Y model of the antibody molecule, in which the stem is formed by the carboxy-terminal half of the two heavy chains whereas the two arms are formed by the amino-terminal halves of the heavy chains paired with the entire light chains. One antibody molecule has two antigen attachment sites located on the two arms, and the corresponding proteolytic fragments were called Fab fragments (ab = antigen binding). The proteolytic fragment corresponding to the stem happened to be easily crystallizable and was called Fc (c = crystallizable). It exerts generic functions such as, among others, binding to Fc receptors on cell surfaces. In addition, several different types of light and heavy chains could be distinguished. The latter led to the distinction of a handful of immunoglobulin (Ig) classes such as IgM, IgG, IgA, IgE, etc., which serve different roles in immune defense mechanisms.

While many immunologists adhered to the instructional theory of antibody diversity well into the 1950s, progress in the knowledge on protein chemistry, of antibodies but also of proteins in general, had begun to seriously undermine the concept of a small number of antibody polypeptides, each capable of folding into multiple conformations. What resulted was a

The immune system, pre-network paradigms

true crisis in the Kuhnian sense, in which a host of new findings, little by little, compromised the credibility of the instruction paradigm until it could no longer be maintained. The revolution commenced in 1955 with a paper by Niels Jerne, proposing what he called the "natural selection theory" of the immune system[17]. Jerne, of Danish origin but leading an international life from his youth[18], had unsuccessfully tried to get himself interested in a number of non-academic and academic disciplines until he had studied medicine in Copenhagen and graduated, at the age of 40, with a thesis on the subject which was to become his lifelong obsession, antibodies[19]. During his thesis work at the Statens Serum Institute in Copenhagen he had studied the role of antibody affinity in the neutralization of toxins, a systematic and sound piece of work without spectacular results. Jerne's breakthrough came during the four post-doctoral years which he spent at the Serum Institute, when he studied antibodies against bacteriophage, which are small viruses which infect bacteria. This was the high time of bacterial genetics, which took advantage of bacteriophage to transfer genes from one bacterial strain to the other, enabling the scientists to establish rather precise maps of bacterial chromosomes. Bacteriophage were thus a hot subject and their properties were studied – among other approaches – by making antibodies, one of the most common tests being neutralization of infectivity. Jerne studied the influence of antibody affinity on phage neutralization by experiments which involved long dilution series of the antisera to be tested, so as to estimate the quantity of antibodies by the highest dilution that neutralized the phage. The sera from non-immunized animals, so-called pre-immune sera, were used as controls. Much to his chagrin, Jerne found that pre-immune sera, i.e. sera from animals that had not been immunized, contained antibodies that could neutralize bacteriophage, albeit in small quantities[20].

Antibodies to bacteriophage were thus not induced *de novo* by immunization, they were only augmented from small to large quantities. The critical piece of luck that made Jerne's discovery possible was that bacteriophage can be highly sensitive to antibodies and may be neutralized by miniscule amounts of antibodies. The amounts of antibodies to a defined antigen in normal serum are so small that conventional assays, such as hemagglutination or precipitation, usually did not detect them, they are below the limit of detection of such assays. The unusual sensitivity of bacteriophage to neutralization by antibodies will be important again later in the context of the network theory where this led investigators, unlike it did for Jerne, on a wrong track (see chapter 10 and 14, interview with Klaus Rajewsky). To Jerne, the presence of specific antibodies to a fortuitously chosen antigen suggested that antibodies of defined specificity must be preformed in the body. Jerne published his "natural selection theory of antibody formation"[17] in the *Proceedings of the National Academy of Sciences of the USA*, then perhaps the most prestigious scientific journal. The natural selection theory postulated that antibody polypeptide chains are preformed in the body in a multitude of different amino acid sequences. It was not known at the time in which ways the amino acid sequence of antibodies determined their specificity, but knowledge on protein chemistry dictated that proteins that bind different ligands must differ in amino acid sequence. Upon entry of an antigen, antibodies that had complementarity to it would bind the antigen and form antigen-antibody complexes in the serum. This initiated a process of multiplication which, according to Jerne's speculation, could involve macrophages that first internalized the complexes and then multiplied the antibody. The exact nature of this process remained undefined in the paper, probably contributing to the vehement rejection that Jerne's theory initially received. Prominent biologists including Felix Haurovitz, Joshua Lederberg, and Linus Pauling, all supporters of the instructional theory, criticized Jerne not only because of the elusive multiplication mechanism but also because he had neglected to quote Ehrlich, who had first postulated a multitude of preformed antibodies[6].

Jernes's paper added to the crisis but did not cause paradigm shift. This did not occur until two years later when Macfarlane Burnet in Melbourne, Australia, augmented Jerne's theory by filling in the missing mechanism of multiplication[21]. In two publications, of 1957 and 1959, Burnet suggested that preformed antibodies exist as receptors on antibody-producing cells[21,22]. Following the binding of an antigen, the cells would be stimulated to divide and to secrete the antibody into the body fluids. The critical difference to Ehrlich's side chain theory was that each antibody-producing cell synthesized only one type of antibody, elegantly explaining that immune responses selectively augment specific antibodies only. The immune system thus consisted of multiple clones of antibody-producing cells, each synthesizing a different antibody, likely resulting from somatic mutations of a few inherited antibody genes in these cells. Immune responses to antigens occur by "clonal selection". Although Burnet's clonal selection theory left many questions open, it was sufficiently consistent and convincing to cause paradigm shift. This was completed in 1959 when Lederberg[23] admitted in an article in *Science* that clonal selection was likely to be the correct explanation for the formation of specific antibodies. He was supported in the same issue by David Talmage[24], who had been on Burnet's side all along.

The clonal selection paradigm was very productive, as it supplied immunologists with a variety of research avenues to follow, of gaps to fill, of experimental systems to establish, of apparatus and methodology to be invented. Indeed under its influence immunology developed into one of the most productive biological sciences, for the first time combining previously separate disciplines such as biochemistry, physiology, cell biology, developmental biology, genetics, anatomy, and clinical medicine. In this way immunology became the first interdisciplinary science, a forerunner of today's more broadly defined life science. Owing to this interdisciplinary, system-oriented approach, many of the breathtaking advances in general cell biology and genetics of the late 20th century were achieved in immunological experiments. Until today many traditionally minded scientists, experts in their academic discipline but purposely ignorant beyond its boundaries, look with suspicion and awe at the success of immunology in the second half of the 20th century.

One field of research opened by the clonal selection paradigm was the question which cells were responsible for recognition of antigens and immune responses, including the production of antibodies. As early as 1945 Merill Chase at Rockefeller University had shown that cells from lymphoid organs could adoptively transfer skin reactivity to tuberculin[25], and in 1953 Avrion Mitchison in Britain had obtained similar results for transplantation immunity[26], both reactions then believed to be due to antibodies. In 1962 James L. Gowans and coworkers in Britain reported that immune responses were initiated by the small lymphocytes that reside in these organs and recirculate through the lymphoid system[27]. Astrid Fragraeus, a young Swedish investigator, had observed that another cell type, called plasma cells, accumulated in lymphoid organs after immunization[28], and Albert Coons and coworkers at Harvard University showed that they were full of antibody molecules[29]. Subsequently, Gustav Nossal in Australia demonstrated in elegant experiments with single plasma cells that each cell produced only one type of antibody with a single specificity[30], again in full agreement with the expectations of the clonal selection theory. These experiments suggested that upon exposure to an antigen, antibody receptors of some small lymphocytes bind the antigen, followed by cell activation, increase in cell size, proliferation, and differentiation into a plasma cell. The plasma cell, as shown by a dramatic enlargement of its endoplasmic reticulum, is engaged in massive protein synthesis. Most of this is indeed newly synthesized antibody, which is secreted into the blood and body fluids. Plasma cells are thus terminally differentiated antibody-producing cells secreting the specific antibodies that arise in the blood in response to an antigen. Antigens are eliminated from the body by different mechanisms depending on the class of antibody. For example, certain IgG antibodies bind to Fc receptors on macrophages which then phagocytize the antigen-antibody complexes and digest them. IgM antibodies activate complement to lyse bacteria, and so on. These observations were gratifying as they suggested that specific immunity could be fully comprehended within the clonal selection paradigm.

An experience that had been made very early in the history of immunology was that the immune system is tolerant of self, i.e. it normally does not make immune reactions to any of the cells and molecules that belong to the body itself. Paul Ehrlich had coined the term "horror autotoxicus", meaning that production of antibodies against self antigens is in principle impossible[31]. However, that the immune system can indeed turn against its own bearer was first demonstrated by J. Donath and Landsteiner, who observed that a certain type of hemolytic anemia was caused by an auto-antibody[32]. The Donath-Landsteiner antibody was the first auto-antibody to be described and the disease is since termed "autoimmune hemolytic anemia". Today we know many autoimmune diseases, often associated with very severe pathology, organ dysfunction, and massive tissue destruction. Self tolerance is under genetic control, as shown by the fact that identical twins accept each other's tissue grafts, while non-identical twins do not. Ever since the clonal selection theory had postulated that the diverse repertoire of antibodies is preformed by random gene mutations in the antibody-producing cells, it had created a problem that remained with immunologists until today, the problem of self – nonself discrimination: How does a highly diverse set of specific antibody combining sites – estimates reached numbers up to 10^{12} different antibody molecules that the immune system can produce – avoid reacting to the very large variety of self molecules whilst it reliably reacts to the even larger variety of foreign structures. As early as 1949 Burnet and Frank Fenner had hypothesized that tolerance of self be induced early in ontogeny by inactivation self-reactive antibody-producing cells[33]. This seemed to be confirmed in 1953 when Peter Medawar and colleagues obtained results in seeming agreement with Burnet's and Fenner's suggestion[34]. They used genetically different mouse strains, which normally reject each other's skin grafts. Injection of cells derived from newborn mice of one strain into newborn recipient mice of the other, however, induced tolerance so that later tissue grafts of that strain were accepted even by the adult mice. Injection of newborn or adult cells in adult mice did not induce tolerance. Based on Medawar's experiment, Burnet[22] as well as Lederberg[23] independently suggested similar mechanisms for tolerance induction. They postulated that clones of antibody-producing cells in newborn animals pass through a period during which they cannot be activated but are rather inactivated or eliminated by the antigens they encounter. Clones producing antibodies with specificity for self structures, which Burnet called "forbidden clones", are silenced during that phase while clones that do not react with self structures remain unaffected. Burnet's and Lederberg's elegant interpretations of Medawar's transplantation experiments received support from experiments concerning antibody production to various other antigens. For example, William Weigle had shown that tolerance could be induced by injection of rabbits with soluble antigens such as bovine serum albumin[35], and David Talmage had reported on induced tolerance to red blood cells[36]. All of these results were interpreted as reflecting inactivation of antibody-producing clones during an early sensitive phase in their generation, in full agreement with the clonal selection theory.

Perhaps the most important prediction of the clonal selection theory was that antibodies of different specificities differ in their amino acid sequence. Although techniques required for determining the amino acid sequences of polypeptide chains had been established, they were still cumbersome, time consuming, and required large quantities of the protein under study. Specific antibodies could not be readily purified from immune sera and certainly not in large amounts, so that this prediction could not be easily tested. However, several investigators including Henry Kunkel at Rockefeller University, New York, had observed that patients with multiple myeloma had large quantities of unusual proteins in their sera which were very similar to immunoglobulins[37]. The tumors were soon identified as plasmacytomas, i.e. malignant counterparts of normal plasma cells. Spontaneously arising from a single transformed plasma cell, each plasmacytoma is a monoclonal tumor and secretes a single type of immunoglobulin, corresponding to an antibody with an unknown antigen binding specificity[38]. Certain plasmacytomas secrete only light chains, so-called Bence-Jones proteins, which are secreted in the urine of the patient and can therefore be easily collected and purified. Reasoning that Bence-Jones proteins of two different myeloma patients almost certainly derive from antibodies with different specificities, Norbert Hilschmann, a German postdoc in the laboratory of Kunkel's colleague Lyman Craig at Rockefeller University, determined the amino acid sequences of two such light chains. He found that they differed in a large number of amino acid positions, with all but one of the differences in the amino-terminal half of the chain[39]. Published in 1965, this corresponded strikingly to the expectations of the clonal selection theory, confirming that different clones of plasma cells secrete antibodies of different amino acid sequence. Moreover, it was the first evidence for the fact, not much later extended to heavy chains as well, that antibody polypeptide chains posses an amino-terminal variable (V-) region and a carboxy-terminal constant (C-) region. The V-regions of heavy and light chain together are responsible for the antigen binding specificity. The C-regions of the heavy chain differ between the immunoglobulin classes, thus determining the generic functions of the antibody.

While all of these findings nicely fulfilled the expectations of the clonal selection paradigm, other lines of research yielded results that could not be readily accommodated in that thought style. Most of this additional evidence was not so much at variance with clonal selection but was simply not covered by it, it came unexpected and unforeseen. In addition, thinking about immunity in terms of clonal selection created certain problems for which the paradigm did not offer testable predictions. These problems, causing immunology to search for a new paradigm, will be summarized in the following chapter.

Chapter 7

The necessity for an
interactive theory of immunity

Assimilation of a new sort of fact demands a more than additive adjustment of theory, and until that adjustment is completed – until the scientist has learned to see nature in a different way – the new fact is not quite a scientific fact at all.

T. S. Kuhn,
The Structure of Scientific Revolutions

It was known for a long time that certain immune reactions, particularly some that manifested themselves as skin swellings, were elicited by cells rather than soluble antibodies. For example, classical experiments of Chase in 1945 had shown that immunity of guinea pigs to tubercle bacilli, as tested by a skin reaction to tuberculin, was transferable to non-immune animals by lymph node cells but not by serum of immune animals[1]. This type of immunity had been termed "cellular immunity", not knowing what it really was, to distinguish it from antibody-mediated "humoral immunity". In his first paper on the transfer of transplantation immunity by lymph node cells in 1953 Mitchison commented "transplantation immunity shares with … immunity to tuberculin the property of being transferred with greater facility by cells than by serum"[2]. In the absence of a paradigm that included immune mechanisms without antibodies, these observations were not followed by any sort of interpretation.

Before 1961 the thymus was not considered to have immunological functions, even though it contained massive amounts of lymphocytes and only a few non-lymphoid cells. Thymectomies had been performed on humans for therapeutic reasons without significant adverse effects on their health. A number of investigators had surgically removed the thymus from adult mice, without any noticeable impairment of immune responsiveness thereafter. Moreover, unlike lymphocytes derived from lymph nodes or

The necessity for an interactive theory of immunity

thoracic duct of immunized mice, lymphocytes derived from the thymus of immune mice did not transfer immunity to non-immunized recipient mice. This fact led Peter Medawar, British guru in immunology because of his famous experiments in transplantation immunity, to conclude that "we shall come to regard the presence of lymphocytes in the thymus as an evolutionary accident of no great significance"[3]. Medawar's view was seriously shaken first by studies showing that mice thymectomized as newborns developed a number of immunological abnormalities as adults[4]. For example, they accepted, as adults, skin grafts from a genetically distinct strain and failed to mount proper immune responses to experimental immunization. Moreover, Henry Claman and coworkers of the University of Colorado had reported that, upon transfer of cells from the bone marrow and the thymus into irradiated recipient mice, only the combination but neither cell type alone was able to mount an antibody response[5,6]. These experiments provided the first evidence that two cell types, one derived from the bone marrow and the other from the thymus, were required for antibody responses. The reputation of the thymus as a pivotal immunological organ became generally accepted when Jacques Miller and Graham Mitchell in Australia showed that recovery of immunocompetence after irradiation required an intact thymus, whereas thymectomized mice that had been irradiated never recovered immunocompetence even if they were reconstituted with bone marrow[7–9]. Their experiments provided the foundation for what is now a solid scientific fact, proven time and again and unchallenged until today, namely that antibody production by bone marrow-derived (B-)lymphocytes requires the help of thymus-derived (T-)lymphocytes. The concept of cell cooperation in the immune system was born. Burnet was highly irritated by Miller's and Mitchell's papers, published in 1967/8, and aggressively expressed his contempt about the significance of results obtained in such "biological monstrosities as inbred mice that were thymectomized, lethally irradiated, and protected by injection of bone marrow from another mouse"[3]. Obviously, Burnet's irritation resulted from the realization that, if Miller's and Mitchell's results were indeed correct, the clonal selection theory needed significant revision.

Rapid progress was also made in the unravelling of the differentiation pathways of lymphocytes. The first lymphocyte differentiation antigen, theta, was described in 1969 by Martin Raff in Britain[10]. Using the novel technique of immunofluorescence[11], Raff demonstrated that theta was expressed on the surface of all T cells whereas B cells had immunoglobulins on their surface, with no overlap between the two types of cells[12,13]. This allowed for the first time to study the different developmental pathways by which T and B cells are generated. Like all cell types in the blood, both types of lymphocytes arise by differentiation from hematopoietic stem cells that reside in the bone marrow. Via several partially committed intermediate progenitor stages they finally differentiate into mature small lymphocytes. B cells complete their entire development in the bone marrow. T cells leave the bone marrow as immature precursors and undergo final maturation in the thymus. In a constant process throughout life, mature T and B cells leave their tissues of origin, i.e. the thymus and the bone marrow, respectively, travelling to the peripheral lymphoid organs, such as spleen and lymph nodes, where they settle in well defined anatomical compartments as small resting, i.e. non-dividing cells waiting for their appropriate antigens to come along. They also recirculate, i.e. they travel from these organs into the tissues and then *via* the lymph stream into the blood and back to the lymphoid organs[14–16].

The vast majority of antigens, so-called thymus-dependent antigens, require both T and B cells in order to stimulate antibody production[17,18]. While the B cells produce the antibody, the T cells provide a helper function. The typical antigens used in T cell-B cell cooperation experiments were complexes of a protein, the carrier, and a small chemically attached determinant, the hapten. While the B cell produced antibody to the hapten determinant, the T cell had specificity to the carrier[19,20]. A regular finding was the requirement for hapten-carrier linkage, i.e. immunization with hapten and carrier in non-linked form did not lead to antibody production by the B cell[20]. As a result, it was envisaged that the antigen served as a bridge between the B cell and the helper T cell, allowing the helper T cell to closely interact with the B cell. The phenomenon of T-B cooperation, which took many years to even partially elucidate, provided a paradigm on its own that guided research activities for many years in many laboratories, not covered by the clonal selection theory.

Another cell type, the macrophage, was first described around 1880 by Ilya Metchnikow[21], who shared the 1908 Nobel Prize with Ehrlich. Macrophages also develop from hematopoietic stem cells in the bone marrow, but split off the lymphoid lineages rather early, to form the myeloid lineage. Depending on the organ of residence in the body and maturation state macrophages are called different names, i.e. Kupffer cells, Langerhans cells, etc. Metchnikow had accurately described their role in non-specific immunity, i.e. they non-specifically phagocytize particulate antigens such as bacteria. For many years it was known that when the antigen is covered with antibody, phagocytosis is strongly enhanced, a process termed opsonization[22]. In addition, some experiments had already shown that macrophages seemed to be necessary to induce helper T cell-dependent antibody responses in tissue culture[23,24]. An alternative model of T-B cooperation envisaged the macrophage as a third cell type, presenting the antigen on its surface for recognition by both T and B cells[25]. The macrophage thus functioned as an "antigen-presenting cell". While macrophages had not played much of a role in theories of adaptive immunity until then, they had suddenly entered the stage as a third cell type participating in specific immune responses.

Helper T cells were found to possess immunological specificity. Indeed, their potency of discrimination among foreign antigens in T-B cooperation experiments was of a similar magnitude as that of B cells, consistent with a clonal expression of highly specific receptors. This presented immunologists with a long standing problem, because no matter how hard one looked, T cells did not produce antibodies. There were different ways in which immunologists dealt with this problem, which will be discussed in the context of the network theory (see chapter 10). Suffice it to say at present that the clonal selection theory did not provide a framework how to deal with this problem. Moreover, whereas initial transplantation experiments *in vivo* had assumed a critical role of antibodies and B cells in transplant rejection, it became gradually obvious that the cells mediating transplant rejection were T cells. An important role in transplantation immunology played the mixed lymphocyte reaction, in which recipient lymphocytes were mixed with irradiated donor cells in a test tube in order to assess the degree of histocompatibility[26,27]. The recipient lymphocytes would react against the cells of the donor by more or less vigorous proliferation, which could be measured as a parameter predicting the acceptance or rejection of a transplant. The cells proliferating in mixed lymphocyte reaction were almost exclusively T cells. Much to the surprise of immunologists, the proportions of recipient T cells reactive with the cells of a donor could reach several percent. Thus, while T cells showed exquisite specificity for carrier determinants in T-B cooperation, the tissue antigens that played a role in transplantation immunity were recognized by high proportions of T cells, rather consistent with a degenerate receptor repertoire or with the expression of more than one specific receptor per T cell. A much discussed experiment was contributed by Morton Simonsen in Denmark, who cultured chicken lymphocytes on the chorioallantois of eggs of unrelated chicken[28]. Lymphocytes reacting to the transplantation antigens of the chorioallantois would proliferate and become visible as colonies that could easily be counted. Depending on the combination of lymphocyte donor and egg, the frequency of reactive cells could reach 2%. The reactive cells in Simonsen's experiment were initially believed to be antibody-producing cells, so that the result was discussed as a challenge to clonal selection[29]. Once it was realized that the reacting cells had been T cells, the experiment added to the difficulties of interpreting T cell specificities. The reaction of high proportions of T cells to allogeneic transplantation antigens presented an unsolved problem to immunologists for many years, with no help from the clonal selection theory.

Was there perhaps something special to the antigens involved in transplant rejection, to which T cells reacted in such high proportions? It had long been known that they were products of so-called histocompatibility genes organized at different chromosomal loci. While a large number of histocompatibility loci were identified in mouse and man, only one of them controlled fast and vigorous graft rejection, therefore termed the major histocompatibility complex (MHC)[30]. At the time the analysis of the MHC largely relied, besides conventional Mendelian genetics, on serology and the mixed lymphocyte reaction[31], leading to the definition of two types of antigens, serologically defined (SD, later called Class I) and lymphocyte defined (LD, later called Class II) antigens. Both groups of antigens were encoded by a small number of linked genes each, all occurring in large numbers of allelic variants in the population[32–34]. Each individual had two allelic forms of each of these genes. Because of the pronounced polymorphism, the probability for two unrelated individuals to have the same allelic composition was negligible. This knowledge merely aggravated the problem posed by the high proportions of T cells reactive to foreign MHC antigens: If there were hundreds or perhaps thousands of different MHC alleles in one species, how can 2% or more of the T cells of one individual react with the MHC antigens of another? If MHC recognition were specific, a proportion of 2% would be compatible with merely 50 different MHC types in the population. This apparent degeneracy of MHC reactivity contrasted sharply with the fine-tuned specificity of helper T cells.

A novel type of T cell, the killer T cell, was discovered by Pekka Häyry in Finland in 1970[35,36], and subsequently analyzed in great detail by Jean-Charles Cerottini and Ted Brunner[37] in Lausanne, and by many others later. While until then it was believed that T cells had only cooperating functions in immune responses but no role as effectors in immune defense mechanisms, killer T cells were now shown to be able to kill target cells, such as allogeneic cells or tumor cells, by a process of cell lysis, without the participation of antibodies. The discovery of T cell-mediated lysis was quite revolutionary because until then all destructive functions of the immune system, from bacterial killing to transplant rejection, had been thought to be associated with antibodies. Now it became possible to consider that T cells alone would be responsible for transplant rejection, in line with the observations in mixed lymphocyte reactions. Killer T cells recognizing allogeneic MHC antigens were easily generated in mixed lymphocyte reaction[38,39], in line with a prominent role in transplant rejection. In addition, killer T cells with specificity for virus-infected cells could be recovered from virus-infected mice[40]. T cells were thus, besides their helper function for B cells, also directly involved in various immune defense mechanisms. T cells therefore seemed to comprise several subpopulations, which were not only different in their functions but could also be distinguished by surface markers with the help of specific antisera, at the time still produced by conventional immunization. After the discovery of the theta antigen, the marker for all T cells, soon followed the Ly-series that allowed to differentiate helper and killer T cells[15]. Together with the B cells, three types of antigen-specific lymphocytes could now be distinguished. The cells participating in immune reactivities thus presented themselves in an ever increasing complexity, far beyond the simple concept of antibody-producing cell clones reacting to antigens.

Another line of research hinting at an important role of genetic elements in regulating immune reactivities was concerned with the so-called immune response (Ir) genes. Ir genes controlled the strengths of antibody

The necessity for an interactive theory of immunity

responses to certain model antigens in mice such that one strain of mice produced high levels and another strain produced low levels of antibodies to the same antigen[41,42]. In breeding experiments, high and low responsiveness were heritable according to Mendelian rules. Other than expected, these Ir genes did not control B cell responsiveness but rather that of helper T cells and, much to everyone's surprise, mapped to the MHC complex, as first observed in guinea pigs[43]. Soon enough, certain killer T cell responses were also found to be under genetic control and the Ir genes involved in the responses of killer T cells also mapped to the MHC. While helper T cell-dependent antibody responses seemed to be controlled by LD loci, those of killer T cells were related to SD loci. There were extensive speculations on the putative ways in which the MHC could possibly determine the responses of T cells to foreign antigens that bore no apparent relation to MHC antigens.

If the reader finds this account of the knowledge on T cells around 1970 confusing, be assured that it was no less confusing to the immunologists of the time. It thus came as a relief when Rolf Zinkernagel and Peter Doherty reported the phenomenon of MHC-restricted antigen recognition of T cells in 1974[44], earning them the Nobel Prize in 1996. They had found, studying killer T cell immune responses to virus infections in mice, that the induced virus-specific killer T cells lysed target cells only when they (i) were infected with the same virus used for infection of the mice, and (ii) when they stemmed from mice that had the same MHC as the infected mice. In due course, corresponding results were obtained for T cells involved in cell-mediated immunity and antibody responses[45–47]. For example, helper T cells cooperated only with syngeneic B cells, i.e. B cells that shared their MHC. Killer T cells required sharing of Class I antigens whereas helper T cells required sharing of Class II antigens. The conclusion was that T cells possessed two specificities, one for the antigen and the other for syngeneic MHC molecules. When they killed a target cell or helped a B cell, this interaction required, in addition to the foreign antigen, the presence of the correct, syngeneic MHC molecules. The T cell thus had two specificities, which were born either by one receptor specific for MHC molecules that were somehow modified by the antigen, or by two receptors, one specific for MHC and the other for the antigen. These observations opened new avenues for interpretations and experiments to resolve the long standing problem of the high proportions of MHC-reactive T cells. In addition, and most intriguingly, an element of self-recognition had been introduced, an apparent paradox to the established role of the immune system to react to foreign structures. Again, the clonal selection paradigm had no solution to offer.

Not only clonal selection by antigen, also Burnet's and Lederberg's theories on self-tolerance by inactivation of B cell clones were in bad shape when a number of older and younger discoveries were considered together. Neonatally thymectomized mice accepted allogeneic skin grafts[3].

Neonatally induced transplantation tolerance needed reinterpretation, as transplant rejection was caused by T cells rather than antibody-producing B cells. In antibody responses, the duration of tolerance induced by injecting soluble antigens was drastically modified in thymectomized mice[48], pointing to a complicated role of T cells in regulating the production of antibodies to self-antigens by B cells. The worst blow to clonal inactivation mechanisms in self-tolerance, however, came from the discovery of suppressor T cells in 1971, independently by Wulf Droege[49] and Richard (Dick) Gershon and coworkers[50], by the latter subsequently expanded into a paradigm in its own right[51,52]. The initial observation was that thymocytes, when admixed to lymphocytes proliferating in response to antigens, could suppress the extent of proliferation. Similar suppressive effects were seen in due course in helper T cell-dependent antibody responses and in assays of cell-mediated immunity. Suppressor T cells were found to be antigen specific, but it was unclear at first which cell type they suppressed, helper T cells or B cells. They represented a specialized subset of T cells as demonstrated by surface markers, and thus added to the complexity of the cellular composition of the lymphoid system. They provided an element of symmetry, suggesting that the strength of immune reactions was the result of a balance of helper and suppressor T cells. In other words, suppressor T cells offered themselves for theories on immune regulation and self-tolerance, they seemed to be the missing link by which the control of all immune reactions could be attributed to positively and negatively regulating T cells. Consequently, many immunologists developed the feeling that the entire functional behavior of the immune system, including self-tolerance, was coordinated by T cell subsets.

Obviously, a new theory on acquired immunity and self-nonself discrimination was needed. There were two early attempts at comprehensive theories, one by Jerne[53] and the other by Peter Bretscher and Melvin Cohn[54]. Both were launched within a few months of each other in 1970/71, i.e. before MHC-restricted antigen recognition of T cells and suppressor T cells had been described, so that both were soon outdated in their original versions. However, they will be briefly outlined here, as they disclose some of the differences in thought styles between the chief proponent-to-be of the network theory, Jerne, and his chief opponent-to-be, Cohn.

In his theory "The somatic generation of immune recognition"[53], published in January 1971 as the first paper in the first issue of the newly founded *European Journal of Immunology*, Jerne expanded Burnet's clonal selection theory by further elaborating on the generation of antibody diversity by somatic mutation of antibody genes, and incorporating the predominant reactivity of the immune system to allogeneic MHC as a novel aspect. Important roles in the theory were also played by the control of antibody responses by MHC-linked Ir genes, and the fact that pronounced spontaneous lymphocyte proliferation took place in certain lymphoid tissues, most prominently the thymus. Key to the theory was the novel hypothesis that the inherited antibody genes in the genome, then termed "germline" genes, encoded antibodies with specificities for all the MHC antigens of the species. Because an individual expresses only a small proportion of the MHC antigens of the species, the newly arising B cell clones of an individual would consist of two subsets, a smaller one reactive with the individual's own MHC antigens, and a larger one reactive with all the other ("allo"-) MHC antigens of the species. The allo-MHC-reactive set contains the clones that react with the MHC antigens of other individuals of the species, for example upon tissue transplantation. In this way, the majority of clones in the immune system exhibits allo-MHC-specificity, thus nicely explaining the predominant recognition of allogeneic MHC antigens by the immune system as due to B cell clones which express non-mutated "germline" antibody genes. The self-MHC-reactive clones, in contrast, encounter their antigens as they arise and, in the sense of Burnet's forbidden clones, become inactivated as long as they maintain self-reactivity. They can escape inactivation, however, if they mutate their antibody genes so as to lose self-reactivity. These mutations take place upon proliferation in the lymphoid organs which Jerne viewed as "mutant breeding organs". Mutations are per definition random and thus generate a large number of different antibody genes. Maintenance of functionality in antibody genes that have accumulated random mutations is likely a rare event, with many mutations resulting in deterioration of the gene. Thus, the process would usually yield no more than one pair of functional heavy and light chain genes in one B cell. The antibodies encoded by these genes are depleted of MHC specificities but include antibodies with any other specificity for environmental antigens. The proportion of clones reactive with any given foreign antigen would be very small compared to that reactive with allo-MHC.

Jerne's theory, satisfying the postulate of simplicity, explained the strong allo-MHC-reactivity of the immune system, the clonal distribution of unique antibody receptors, and self-tolerance, all by a single novel hypothesis. Elegant as it was, though, this hypothesis turned out to be far off the mark, a nice idea but pure phantasy. This was the main but not the only reason why the theory failed. It claimed to contain a satisfactory explanation for the apparent genetic control of the strength of antibody responses by MHC-linked Ir genes. Since MHC genes determined which section of the inherited antibody genes had to somatically mutate in order to generate antibodies for foreign antigens, no wonder that antibody responses to such antigens differed among strains of mice or guinea pigs with different MHC. Here, however, it showed that Jerne refused to take notice of the existence of T cells which he regarded as no more than a subset of lymphoid cells with somehow elusive antibody receptors. It had long been established that antibody production required the cooperative response to two determinants on the antigen, and Jerne had even coauthored one of the initial papers demonstrating this phenomenon in 1969[19]. The recognition of hapten and carrier determinants by B and T cells, respectively, had been amply demonstrated by 1971, and the possibility that Ir genes controlled the ability or inability of helper T cells to respond to the carrier was a subject of acute discussion. Not addressing any of these new burning problems in immunology, Jerne's theory was in part outdated already before it was launched. Although until to date it has accumulated respectable 787 quotations, it had no lasting influence on thought styles in immunology.

Bretscher's and Cohn's paper entitled "A theory of self-nonself discrimination" appeared in September 1970 in *Science*[54]. Key to their theory was the requirement for the recognition of two linked antigenic determinants in antibody responses to hapten-carrier model antigens. It proposed as a first new hypothesis that in order to be activated, the B cell has to receive two signals, signal 1 from its antibody receptor that has bound the antigen *via* the hapten, and signal 2 from a soluble antibody bound to a carrier determinant. By binding to a carrier determinant, the carrier antibody is connected to the cell surface *via* a bridge formed by the receptor antibody and the antigen. While the receptor antibody is integrated in the B cell membrane, the carrier antibody contacts the cell from the outside. The nature of the signals delivered to the cell was not further specified, except that they differed one from the other, possibly owing to different immunoglobulin classes of the two antibodies. The second new hypothesis in the theory was that if the B cell received only one signal, signal 1, it is not activated but tolerized. Newly arising B cell clones with specificity for self-antigens will immediately receive signal 1 and will be tolerized. Newly arising B cell clones with specificity for foreign antigens will not be tolerized and can accumulate. When a foreign antigen with two or more determinants is introduced, two of such clones cooperate, one binding the antigen *via* its receptor antibody and the other contributing the carrier antibody. Clones recognizing any two determinants of the antigen can cooperate, with the designations hapten and carrier interchangeable. The theory thus offered a plausible mechanism for self tolerance, which differed from that previously proposed by Lederberg and Burnet in that antibody-producing B cells could be tolerized at any point in time, not only as newly arising clones. This was much more in line with many results on induced tolerance in adult animals.

Bretscher and Cohn addressed also the cellular aspects of their theory. Taking into account the recent data on the role of the thymus and the helper effect of T cells in hapten-carrier responses, they considered the possibility that the carrier antibody which provided signal 2 was contributed by the helper T cell, which later turned out to be in the right direction. While in this respect the theory was superior to Jerne's, it was not better as far as Ir gene control and the role of MHC was concerned. The predominant reactivity of the immune system to MHC was not addressed. Moreover, although the MHC linkage of Ir genes had been described in several experimental systems, the theory ignored this, accepting that "non-responders possess no carrier antibody" as a fact of life. In one respect, however, Bretscher's and Cohn's theory was clearly superior to Jerne's. It provided ample opportunity for adjustment to novel experimental findings by making auxiliary hypotheses[55]. So the nature of signal 2, which in the original version was delivered by carrier antibody, changed rather quickly to helper/inducer T cells, and later to their cytokines and co-stimulatory molecules. In this way, under the terms "two signal theory" or "associative recognition theory", it survived for many years to come. In its various forms adjusted to novel results it accumulated over 1,000 quotations over the years and became the main and only competitor of the network theory (see chapter 9).

Chapter 8

Proto-ideas of the network theory: antibody self-regulation, idiotypy, the brain analogy, and cybernetics

> *Proto-ideas must be regarded as developmental rudiments of modern theories and as originating from a socio-cognitive foundation.*
>
> Ludwik Fleck,
> *Genesis and Development of a Scientific Fact*

According to Ludwik Fleck, novel scientific knowledge, such as theories or factual connections, often originates from ideas that have been around for a long time[1]. The German term *Uridee* used by Fleck is only poorly translated as proto-idea or pre-idea. Fleck developed the notion of proto-idea using the Wassermann reaction as an example, which exploited alterations in the blood of syphilitic patients as a diagnostic principle. Changes in the composition of blood had been envisaged in relation to syphilis already in the middle ages, and were given names such as "befouled blood", or "hot and thick blood" having a "poisonous" or a "mysteriously bad quality" and being in an "unnatural state". Over the centuries many attempts were made to define theses changes in chemical or microscopic terms, all resulting in claims soon to be revealed as false. For example, in 1872 a Dr. Lustbader reported at a meeting of the society of physicians in Vienna the discovery of "syphilitic particles" in the blood of patients, a finding that was quickly retracted as non-specific. According to Fleck, the establishment of the Wassermann reaction was merely another attempt to verify the existence of chemical changes in syphilitic blood, this time successful, but with no more insight into the nature of the alterations than before[1,2].

Proto-ideas that give rise to and are incorporated into novel scientific theories often originate from within a scientific discipline and are carried along, continually adjusted to its changing thought styles. I will refer to them as esoteric proto-ideas, as they belong to the esoteric circle that shares a though style, the thought collective. In addition, elements of unrelated thought styles may be newly adopted into a scientific discipline and may then serve as proto-ideas in the development of a new thought style. Indeed, the incorporation of paradigms of unrelated disciplines is a main force in the transformation of thought styles that, according to Fleck, is the essence of progress in science. Paradigms of unrelated disciplines, however, are often not fully understood by the members of the thought collective, who are laymen in the unrelated discipline. In order to appear attractive, a paradigm of an unrelated discipline must open new perspectives of thinking, an attribute more readily perceived if constraints of definition are unknown and can be ignored. The form in which paradigms are adopted by another discipline is therefore often more like the fuzzy conception that a scientific fact enjoys in the general public. Paradigms adopted across discipline borders thus often lose the precision of definition, and change their meaning, they become proto-ideas. Such proto-ideas will be referred to as exoteric, with the understanding that they are exoteric only with reference to the discipline adopting them. The change in meaning may be so drastic that little resemblance remains to the original. While there have certainly been more, I will introduce only four of the more obvious proto-ideas of the idiotypic network theory, two of esoteric and two of exoteric origin.

The first of the esoteric proto-ideas to be discussed is the notion of antibody self-regulation. Already in 1961 Jonathan Uhr and coworkers had observed that experimental animals, when injected with antibodies prior to immunization with an antigen, produced lesser amounts of antibodies than animals that had not been pretreated with antibody[3]. Passively administered antibodies thus suppressed antibody formation, suggesting that the strength of the antibody response was regulated by antibody itself, a form of self-regulation. The phenomenon was soon reproduced in many laboratories and investigated in much detail, representing a small paradigm in immunology of the 1960s. It had just become possible to separate two major classes of immunoglobulins, IgM and IgG, on the basis of their difference in molecular mass resulting in different Svedberg constants in the ultracentrifuge, 19S and 7S, respectively[4]. Upon injection of an antigen, 19S IgM was produced first, followed by 7S IgG. Examined for their abilities to suppress upon passive administration, 7S was found to be very effective in suppression, whereas 19S was either weakly suppressive[5] or even enhanced the antibody response under certain experimental conditions[6]. The terms 7S inhibition and 19S enhancement were born, conveniently providing negative and positive feedback mechanisms to be invoked for immune regulation. This made good biological sense, suggesting an initial positive and a subsequent negative feedback during antibody responses, in line with the standard kinetics of antibody responses to many antigens[7].

The molecular mechanisms responsible for 7S inhibition and 19S enhancement were subject to much discussion. Most favored for 7S inhibition were versions based on the competition for the antigen between soluble and cell-bound antibodies: Antigen which is already complexed with soluble antibody is unable to reach the cell-bound antibody receptors, which are required for induction of antibody formation. More difficult to envisage were mechanisms for 19S enhancement. Jerne, who studied the phenomenon together with his coworker Claudia Henry[6], considered the possibility that complexes of antigen with IgM would cytophilically attach to macrophages. The high local concentration of antigen thus achieved on the macrophage surface would facilitate interaction with the antibody receptors on lymphocytes. This mechanism seems like a merger between parts of Jerne's natural selection[8] and Burnet's clonal selection[9,10] theories. Jerne had suggested that preformed soluble antibody, meanwhile known to be of IgM class, would capture antigen in solution and these complexes would be internalized by macrophages which would then replicate the antibodies. Clonal selection, on the other hand, had invoked that antigens bind to antibody receptors on specialized antibody-producing cells, thus activating these cells to produce soluble antibodies. 19S enhancement, which was not readily explained by clonal selection, thus offered Jerne a mechanism to rescue the role of soluble antibody in the initiation of the immune response. No matter what, neither mechanism could be verified, and so the matter remained open to alternative interpretations and speculations, as well as to manifold experimental attempts at clarification.

Undoubtedly the most important esoteric proto-idea of the network theory arose from the notion of idiotypy. Two independent avenues of research, one followed in Henry Kunkel's laboratory at Rockefeller University in New York, and the other by Jacques Oudin and coworkers at the Pasteur Institute in Paris, eventually merged to generate the notion of idiotypy, the term for the uniqueness of an antibody as detected by another antibody. Already in the 1950s, Kunkel and coworkers had begun to investigate apparently novel proteins which appeared in large quantities in the sera of patients suffering from multiple myeloma, malignant tumors arising in the bone marrow. One important question was whether these proteins represented pathologically enhanced quantities of one of the several hundreds of normal serum proteins, or were essentially novel, not present in serum of healthy individuals[11,12]. To indicate their unknown nature, myeloma proteins were initially referred to as paraproteins. One hint as to the nature of paraproteins came from electrophoretic analysis, which showed that many paraproteins migrated in the region of immunoglobulins. However, whereas normal immunoglobulins migrated as a broad zone suggesting pronounced heterogeneity, paraproteins migrated as sharp bands indicating molecular homogeneity. For situations like this, in which the rela-

Proto-ideas of the network theory

tionship of an unknown substance to already known substances needed clarification, one of the most powerful technologies available at the time was serology. Investigators had collections of antisera on the shelf which had been prepared against most known substances and which could be tested for reactivity with any new substance that came along, and it was soon clear that most paraproteins reacted with antisera prepared against normal immunoglobulins. In order to corroborate such a one way identification, careful investigators also used the reverse approach, i.e. they prepared an antiserum against the novel substance and checked its reactivity with a panel of known substances. This is what Kunkel did with a large number of paraproteins, the common result being that an antiserum against a paraprotein reacted with the normal immunoglobulin of the same class, but contained additional antibodies that did not react with normal immunoglobulin[13]. These additional antibodies also failed to react with any other paraprotein, they were essentially specific for the paraprotein used for immunization. Thus, paraproteins bore strong resemblance to immunoglobulins, but carried an additional serologic specificity, termed by Kunkel "individual antigenic specificity"[14].

What was the difference between a paraprotein and normal immunoglobulin of the same class leading to the phenomenon of individual antigenic specificity? The answer to this question was found in the molecular homogeneity of paraproteins as opposed to the heterogeneity of normal immunoglobulin. The latter consists of millions of different antibodies which share the same constant regions but differ in the variable regions of their polypeptide chains. In contrast, a myeloma protein is produced by a myeloma tumor, a tumor arising from a single plasma cell secreting a single type of antibody consisting of a single pair of heavy and light chains, each with just one variable region. Accordingly, antisera against a myeloma protein contain two types of antibodies, against the constant and the variable regions. While the antibodies against the constant region react with normal immunoglobulin, the antibodies against the variable region fail to find the few molecules, if any, with identical variable regions among the vast numbers of different antibodies in normal immunoglobulin. Since each myeloma tumor secretes a unique antibody, owing to the random process of malignant transformation, the antibodies against the variable region of one myeloma protein almost never react with another, they define its individual antigenic specificity.

The second avenue of research that gave birth to the notion of idiotypy began with attempts to understand the genetic control of antibody formation. Also in the 1950s Jacques Oudin in Paris observed that rabbits, when immunized with immunoglobulins of another rabbit, can produce antibodies that recognize the immunoglobulins of some but not all rabbits in an outbred colony[15]. This was unexpected because antibody production against an antigen derived from an individual of the same species seemed a violation of self-tolerance. However, the phenomenon of genetic polymorphism was known from many other examples, so that an alternative interpretation was that different rabbits produced structurally different immunoglobulins owing to polymorphisms in the genes encoding immunoglobulin polypeptide chains. A systematic analysis of the reactivities of syngeneic anti-immunoglobulin antisera among systematically bred rabbits soon revealed that positive and negative reactions with a given antiserum were inherited in a Mendelian fashion, suggesting that for each immunoglobulin gene, including heavy and light chain constant and variable regions, several allelic variants existed in the rabbit population. The phenomenon was termed "allotypy", from the Greek αλλοσ (different) and τυποσ (type). Allotypy was found also for human[16] and mouse immunoglobulin genes[17] and was for many years the main, if not the only way to study the genetics of antibody formation. For example, the fact that IgM and IgG antibodies to the same antigen may share the same V region gene product was first suggested by allotype sharing[18].

Oudin went on in his attempts to define antigenic properties of immunoglobulins by immunizing rabbits with specific antibodies derived from allotypically identical rabbits. Ordinarily, since all immunoglobulin genes are the same between two allotypically identical rabbits, no antibody should be produced. Nevertheless, as shown first for rabbit antibodies to *Salmonella typhi*, anti-antibodies were induced, reacting with the antibody used for immunization, but not with any other immunoglobulin preparations including antibodies against salmonella from other rabbits or antibodies to unrelated antigens from the same rabbit[19,20]. These antibodies reacted specifically with the antibodies to a given antigen formed by an individual rabbit. Oudin dubbed the phenomenon "idiotypy", from the Greek ιδιοσ (unique). Antibodies induced by a given antigen shared a set of antigenic determinants, together termed the "idiotype", against which "anti-idiotypic" antibodies could be induced. After some discussion among the experts it was universally agreed that idiotypy and individual antigenic specificity were in principle similar, i.e. they denote antigenic properties of the variable region of an antibody. The term individual antigenic specificity was thereafter dropped and the term idiotypy was used for the phenomenon in all of its experimental variations. The experiments of Oudin, in agreement with earlier studies by Kunkel and coworkers[21], established the notion that not only monoclonal myeloma proteins but also induced antibodies displayed idiotypes. Thus, idiotypy was not limited to homogeneous pathological immunoglobulins produced by monoclonal tumors but applied to all antibodies induced upon normal immune responses and likely of some degree of molecular heterogeneity.

Idiotypy became a major subject of study in many immunological laboratories*. For example, determinations of the frequency of a given idiotype

---

* A comprehensive compilation of idiotypic studies can be found in Greene MI, Nissonof A (eds) (1984) *The Biology of Idiotypes*. Plenum Press, New York.

among normal immunoglobulins were used for estimates of the diversity of antibody variable regions. These experiments, mostly done by titrating the excess of normal immunoglobulin needed for inhibition of the binding of the anti-idiotypic antibody to its idiotype, yielded estimates in the order of one molecule in multiple millions. Attempts to translate such numbers directly into the degree of antibody diversity received criticism because of the unknown extent of cross-reactivity of an anti-idiotypic reagent, i.e. one could not be certain as to what extent it reacted with antibodies whose variable regions were only similar but not identical to the immunizing antibody or myeloma protein.

Another line of research exploited idiotypy for studies on the genetics of antibody formation. Allotypy had been successfully applied to delineate the Mendelian inheritance of antibody constant region genes. Analogously, it was hoped that idiotypy could serve as a tool to analyze variable region gene inheritance, and could thus help answering the pending questions of the numbers of antibody variable region genes in the genome and the putative somatic generation of variable region gene diversity. Specifically, if variable region genes were inherited and expressed by antibody-producing cells in non-mutated form, this should be demonstrable, for example in systematically bred rabbits, by inherited idiotypes of antibodies produced, for example, by parents and offspring. On the other hand, if antibody variable regions were the products of randomly mutated genes in somatic cells, idiotypes should be unique to the antibodies of individual animals and not be heritable. Indeed, the inheritance of idiotypes could be demonstrated in several experimental systems in rabbits and inbred mice[22]. However, these results remained inconclusive for the same reasons as the diversity estimates. It was not possible to prove that the reactivity of parent and offspring antibodies with the same anti-idiotypic antiserum really required structural identity of the variable regions and thus implied the expression of heritable and non-mutated variable region genes. On the contrary, later structural studies showed that most anti-idiotypic antibodies reacted, in addition to the immunizing idiotype, with groups of antibodies whose variable regions did not need to be entirely identical[23], nor did they have to react with the same antigen[24].

A much discussed question was the relationship between the idiotype of an antibody and its combining site for antigen. The phenomenon as such implied that idiotypic determinants must be born by the variable regions of antibodies. Structural studies on antibodies with defined idiotypes confirmed the notion that idiotypic determinants reflected the amino acid sequences of the variable regions of heavy and light chains[23]. In many cases a competition between the binding of the anti-idiotype and of the antigen to the same antibody could be demonstrated[25]. However, this could be explained by either steric hindrance between the two bulky ligands, or by conformational changes across some distance, and thus did not mean that idiotype and antigen binding site were identical. Nevertheless, in certain cases the evidence suggested a pronounced if not a complete overlap between the antigen-binding site of an antibody and its idiotype. In these cases, the antigen and the anti-idiotypic antibody could be envisaged to possess identical molecular surfaces, a conclusion that gave rise to the concept that anti-idiotypes could replace the antigen in vaccination strategies[26]. Moreover, since antibodies can be produced against any antigen, and since anti-idiotypic antibodies could be produced against any antibody, the repertoire of antibody combining sites must include "internal images" of all environmental antigens[27]. The notion of internal image brought a somewhat mystical quality into the discussion of idiotypy, an important factor contributing to the attractivity of the network theory.

A strong element of mysticism was inherent also to the exoteric protoideas of the network theory, most prominently to the brain analogy. The analogy was based on the ability of the immune system to generate a more effective antibody response upon second and subsequent challenges with the same antigen, compared to the first encounter[28]. In other words, the immune system has "learned" to deal with the antigen, based on its ability to "remember" previous exposures to the antigen. Obviously, the terms "learning" and "memory" are borrowed from jargon used in relation to the corresponding properties of the brain. This was possible because the cellular and molecular foundations of learning and memory were poorly understood, even less for the immune system than for the brain, so that the matter lent itself to attractive speculations on putative similarities between the two organ systems.

What does an immunologist know about the changes in the brain associated with learning? The functional units of the brain are the neurons and their connections by axons and synapses. While the number of neurons is constant, the number of axons and synapses is adaptable. A basic set of synapses are genetically determined, they are formed without outside stimuli before birth and are the substrate of a body of inherited knowledge that needs not to be learned. After birth, additional axons and synapses are formed in response to outside stimuli, which are perceived through the sensory organs and their connections to the brain. These novel synapses are connected and disconnected depending on the frequency and strength of activation and reach a steady state which is fixed at around puberty, when the brain "crystallizes", as the neurologists say. The synapses formed from birth to puberty are the substrate of knowledge about the ground rules of life in the world acquired during this time. This knowledge is largely standardized for the members of a given culture but is also to a degree individually variable depending on the body of experiences persons make in their social environment. After puberty, learning is still possible but now it is associated with the frequency of the usage of existing synapses. The more frequently a synapse is used, the higher the quantities of neurotransmitters stored for release upon usage. The knowledge thus acquired is highly specialized and individually variable[29].

Many self-respecting immunologists felt obliged to address the brain analogy in lectures and panel discussions, although little actual research was done on the subject. Indeed, there seem to be some striking similarities between the immune and central nervous systems. Both systems pervade the entire body, and yet function in a coordinated fashion. There is overwhelming complexity, i.e. both systems consist of very large numbers of functional elements, including cells and soluble mediators, which perform their tasks in a quantitatively regulated fashion. As far as learning is concerned, both systems possess the ability to adapt to outside stimuli by undergoing permanent changes in certain elements and/or their interactions, resulting in storage of information, i.e. the accumulation of a stock of memory. The immune system acquires most of its memory between birth and puberty. Thereafter, the thymus involutes and its cellular production slows down dramatically. At least as far as T cells are concerned, the adult immune system has to rely on the existing repertoire to deal with most future challenges. Similarly, most of learned memory in the brain is acquired from birth to puberty, thereafter new synapses are no longer formed and future learning has to rely on using the existing web of connections. All of these properties of the brain, including quantitative regulation, coordinated function, learning, and memory storage, are based on the principle of a multitude of connections between a very large number of functional elements. Was the function of the immune system perhaps also based on connections among its elements? What were these connections, and were they perhaps more profound and widespread than indicated by the merely bilateral T cell/B cell cooperation phenomenon?

Analogies between the brain and computer-operated apparatus were the incentive for the science of cybernetics and its follow-up, systems theory, which will be the second exoteric proto-idea of the network theory discussed here. It started, of all possible activities, with the construction of military equipment. In the 1940s Norbert Wiener, a mathematician at Massachusetts Institute of Technology (MIT), was working on the development of automated goal finding devices for antiaircraft guns, together with Julian Bigelow, a young engineer. The devices were supposed to coordinate the flight pathway taken by the projectile with that of the goal by calculating their trajectories on the basis of multiple measurements including speed, size, gravity, etc., using past recordings of similar events for correction. They built the first computers and succeeded in constructing highly complex devices that recorded a multitude of such parameters, made the corresponding decisions for the aiming process, and translated them into appropriate mechanical action, all in very little time. The success rates increased with time, so that the machines were able to learn from previous recordings of hits and failures, not unlike learning by trial and error by the human brain[30].

Wiener and Bigelow found out that a very important property of a reliable aiming device was a balanced amount of friction in the freedom of movement. If friction was too low, aiming motions often overshot and resulted in uncontrollable oscillations. Wiener consulted with Arturo Rosenblueth, neurophysiologist at MIT, whether similar dysfunctions were known for the human brain. Indeed, Rosenblueth reported that certain injuries to the cerebellum resulted in strongly overshooting and uncontrollable motion of the musculature. As a result, Wiener inferred that mechanical action in both systems was controlled by loops which record the positive impulses and dampen them by negative feedback, if needed. The notion of feedback control was born, a fundamental principle of regulation in complex systems. In addition, complex systems, whatever their nature, seemed to operate according to common principles[31].

Cybernetics not only linked neurophysiology with computer science and engineering, it soon pervaded other fields including mathematics, sociology, economics, anthropology, political sciences, etc.[32]. It generated a tremendous melting pot of ideas, giving rise to concepts such as artificial intelligence, the construction of machines that imitated the brain or other functions of living organisms. The term *system* acquired a meaning of profound generality, comprising complex organizations ranging from human society to microbial ecosystems. In the 1950s the biologist Ludwig van Bertalanffy founded the Society of General Systems Research, which included mathematicians, biologists, biophysicists, economists, etc. It published the General Systems Yearbook, which was to become a most influential periodical for the cybernetic approach to multiple scientific disciplines.

Second-order cybernetics originated around 1970 with an article by Heinz von Foerster entitled "Cybernetics of cybernetics"[33]. Classical (first-order) cybernetics looked at systems as independent of the observer and concerned itself with regulation resulting in steady states and stability, principles applying mainly to engineering and inanimate technologies. In contrast, second-order cybernetics centered on the principle of self-organization that was inherent to living systems of which the observer was an integral part[34]. Thus, second-order cybernetics concerned itself with biological systems, or with societies that have a biological basis. Such living systems are characterized by growth, morphogenesis, and positive feedback loops rather than by homeostasis and negative feedback. They are infinitely more difficult to control, and long term predictions are usually impossible. Notions such as autopoesis, self-production, self-determination, were introduced to denote living systems that are organizationally closed. Prominent theoreticians in second-order cybernetics are, among others, Humberto Maturana and Francesco Varela[35], the latter directly contributing to the conceptual framework of the network theory (see chapter 12). The ultimate consequence of an organizationally closed system is that it includes all the principles steering it, a notion referred to in second-order cybernetics as the "self-reference" of living systems[36]. In other words, "the system collects information about its own functioning, which in turn can influence that functioning; minimal requirements are self-observation, self-reflection, and some degree of freedom of action". The notion of self-reference was originally developed to describe human beings either as individuals or societies. When applied to biological systems of non-conscious nature, it provided just the right touch of mysticism, turning it into an attractive proto-idea for the network theory.

Chapter 9

The idiotypic network theory

Today, Jerne's proposition that such a recondite idiotypic network must regulate the immune system is almost universally accepted.

R.E. Langman and M. Cohn (1986)
The 'complete' idiotypic network is an absurd immune system. *Immunology Today*, 4: 100

Jerne had a talent to present his thoughts in a way that made them appear simple and straightforward. As a sworn reductionist, he looked at the immune system as consisting of lymphocytes and antibodies and began each of his lectures and nearly all of his papers with these numbers: The human immune system consists of 10^{12} lymphocytes and 10^{20} soluble antibody molecules. The number of different antibodies, i.e. the diversity of combining sites which enables the immune system to recognize and discriminate foreign antigens, is not exactly known but was estimated by Jerne to be in the order of five million. The number of each individual antibody molecule, calculated by dividing the total number of antibody molecules by the estimated number of different combining sites, would then be about 2×10^{13}. About 2% of the components of the immune system are turned over every day, i.e. the immune system produces about 2×10^{10} new lymphocytes and 2×10^{18} new antibody molecules every day. The system is thus a dynamic one and requires regulation. For the mouse, the major model animal in experimental immunology, numbers had to be divided by a factor of 3,000, except for the diversity of antibody combining sites which might be of similar size as that of man.

Jerne had invented his own terminology that enabled him to deal with antibody variable regions, antigen binding sites, and idiotypes, etc., in a clear and unambiguous fashion[1]. Three of these terms have been widely adopted by immunologists and will also be used here (see Jerne's own hand-drawings in Figure 9.1): (i) An antigenic determinant, i.e. the part of the molec-

The idiotypic network theory 83

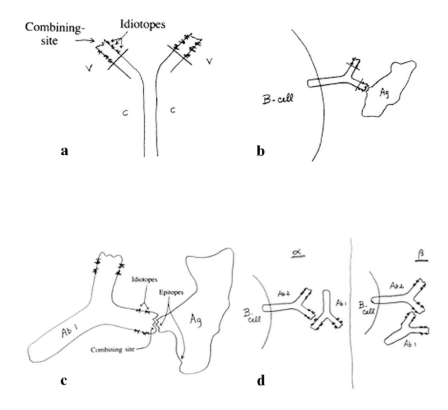

Figure 9.1
Imagery of the immune system à la Niels Jerne
a) A highly stylized drawing of an antibody molecule with linear representation of heavy and light chains. V- and C-regions, as well as the positions of the antigen-combining site and of some idiotopes are indicated. The simplified drawing does not include idiotopes that overlap with the combining site, and ignores the 3-dimensional folding of antibody polypeptide chains.
b) An antibody molecule inserted as receptor in the membrane of a B cell. One of the combining sites is attached to an antigen.
c) A more detailed view of the complex of an antibody (Ab1) with an antigen. The antigen is attached to one of the combining sites of Ab1. The positions of some idiotopes outside of the combining site are indicated.
d) Two ways of Ab1 to interact with B cells expressing Ab2. Left, an idiotope of Ab1 is bound by the combining site of Ab2. Jerne postulated this type of intereaction to be stimulatory for B cells to produce soluble Ab2. Right, Ab1 binds by its combining site to an idiotope of Ab2. This interaction is supposed to be suppressive for the production of Ab2. A third possibility, namely the binding of the combining site of Ab1 to that of Ab2, is not included here. In this case Ab2 would represent an "internal image" of the epitope of the antigen. (From Jerne NK (1985) The generative grammer of the immune system. *EMBO J* 4: 847, © The Nobel Foundation).

ular surface of an antigen to which the antibody binds, is called an *epitope*. An antigen usually has several different epitopes, each of which inducing its own set of antibodies. (ii) The part of the molecular surface of the two variable regions of an antibody binding to the epitope is called the *paratope*. Several different antibodies can have specificity for the same epitope, with a range of affinities depending on the precision of complementarity between epitope and paratopes. (iii) The part of the molecular surface of the antibody variable region to which an anti-idiotypic antibody binds is called an *idiotope*. An idiotope is thus also an epitope, but a special type. A heterogeneous antibody population induced by an antigen represents a collection of idiotypes, and the idiotype of a single antibody molecule consists of several different idiotopes. Some idiotopes may be overlapping with the paratope but there is no stringent association between the two, except that both are localized on the variable region of an antibody.

Idiotypy was originally studied with so-called xenogeneic anti-idiotypic antibodies, produced across species barriers, as in the experiments of Kunkel who immunized rabbits or goats with human antibodies or myeloma proteins[2,3]. Oudin then succeeded in producing so-called allogeneic anti-idiotypic antibodies, i.e. in animals of the same species, by immunizing rabbits with antibodies derived from allotypically matched but otherwise genetically unrelated rabbits[4]. Carrying the game further, a number of laboratories showed that immunization of inbred mice with antibodies derived from mice of the same inbred strain resulted in anti-idiotypic antibodies as well, so-called isogeneic anti-idiotypic antibodies[5,6]. Finally, Scott Rodkey[7] and subsequently others demonstrated that animals immunized with their own antibodies generated autologous anti-idiotypic antibodies. This last experiment closed the argumentative gap by suggesting that idiotypes and anti-idiotypes coexisted within the immune system of an individual, the starting point of Jerne's idiotypic network theory.

The network theory was first launched in 1973[8] when Jerne presented the final lecture of a three-day symposium at the Pasteur Institute in Paris celebrating the 150th anniversary of Louis Pasteur's birth in 1822. He began by giving the audience a tour d'horizon of immunology as a history of "notions", each of which directed thinking in immunology for a period of about 20 years. The first of these were immunization and phagocytosis introduced by Pasteur and Metchnikoff, respectively, in 1870. Via antibodies discovered by von Behring in 1890, cell receptors postulated by Ehrlich in 1900, and antibody specificity delineated by Bordet and Landsteiner in the 1930s, the tour ended with himself and Burnet in the 1950s, with clonal selection. Now, in the 1970s, time called for a new "notion" and he, Jerne, had a clear view in which direction this was going to develop: "This will then lead me to a prediction concerning the theories that will be developed in the period 1970–1990 which, I think, will be of a multicellular type of which attempts to deal with the interaction of T cells and B cells represent only early beginnings".

The idiotypic network theory 85

Indeed, Jerne formulated the idiotypic network theory in this lecture in nearly as much detail as he would ever do in the future. Only minor additions were made in later papers. The concept was based on the established fact that complementary idiotopes and paratopes coexisted in the immune system, with the putative interactions between them resulting in a "formal network" including soluble antibodies as well as lymphocytes that bear antibodies as receptors. He then went on by discussing evidence suggesting that this formal network was indeed suitable to establish a "functional network", stressing mainly three types of conclusions: (i) individual idiotypes and paratopes occur at an average concentration of 10^{10} per ml of blood, consistent with a reasonable probability of mutual encounter, at least as far as soluble antibodies are concerned. (ii) Injection of antibodies to immunoglobulin allotypes[9] and idiotypes[10–12] leads to suppression of the production of antibodies bearing the target allotype or idiotype, showing that antibodies directed against cell receptors can downmodulate the activity of the cells bearing them. (iii) In the absence of external antigens the immune system seems to be in a state of suppression, in part maintained by suppressor T cells[13]. Some suppressor T cells had already been shown to be specific for antibody allotypes[14]. By analogy, Jerne suggested that suppressor T cells may be specific for idiotopes as well. Jerne concluded "that the immune system, even in the absence of antigens that do not belong to the system, must display an eigen-behavior mainly resulting from paratope-idiotope interaction within the system".

As a result, the reactions of the immune system following entry of an antigen are no longer restricted to the stimulation of independent lymphocyte clones bearing appropriately specific receptors, but involve several waves of idiotope-paratope stimulation (Fig. 9.2). The immune response to a single epitope E could thus be described as follows: The first set of antibodies stimulated by epitope E, in later papers termed antibody 1(Ab1)[15], carries a set of paratopes (p1) and a set of idiotopes (i1). Ab1 in turn stimulates Ab2 which is complementary to Ab1 and consists of two subsets: One of these is directed against the i1 idiotopes which completely overlap with p1 so that their paratopes represent an "internal image" of E. The other is directed against i1 idiotopes independent of p1, called the anti-idiotypic set. Both the internal image and the anti-idiotypic sets go on to stimulate Ab3, which again consist of two subsets each, one complementary to p2 and thus mirroring p1, the other directed against paratope-independent idiotopes of Ab2. This would go on indefinitely but as the clonal diversity of each wave increases, the concentration of individual paratopes and idiotopes decreases with successive waves until substimulatory concentrations are reached and the process subsides. Nevertheless, substantial sections of the immune system may be perturbed in responses even to simple antigens.

Jerne postulated a balance between stimulatory and suppressive forces operating in the network: He assumed that a stimulatory signal is generated when a paratope of a cell receptor binds an idiotope (or an epitope),

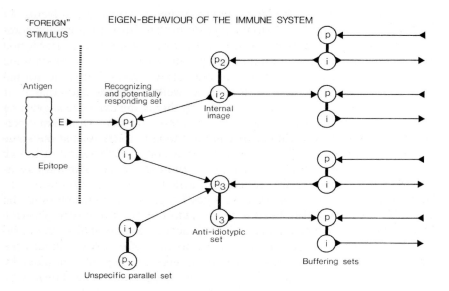

Figure 9.2
Graphic representation of the "Eigenbehavior" of the idiotypic network in response to antigenic stimulation.
An epitope E stimulates antibodies (later termed Ab1) with paratopes p1 and idiotopes i1. These stimulate two sets of anti-idiotypic antibodies (later collectively termed Ab2): An internal image set, with paratopes complementary to p1 and resembling the shape of E, and an anti-idiotypic set, with paratopes complementary to i1 idiotopes. Each of these sets stimulates its own Ab3, again consisting of an internal image and an anti-idiotypic set. The internal image sets cause positive feedback whereas the anti-idiotypic sets cause feedback inhibion, i.e. represent "buffering" sets. Ab2 stimulates, in addition to Ab3, a "parallel set" which shares i1 idiotopes but not p1 paratopes with Ab1. (First published in Jerne NK (1974) Towards a network theory of the immune system. *Ann Immunol (Inst. Pasteur)* 125 C. 373. Jerne Collection, The Royal Library, Copenhagen).

while a suppressive signal is generated when an idiotope of a cell receptor is bound by a paratope (see Fig 9.1). In the absence of external epitopes, because a receptor antibody has multiple idiotopes for each of its two paratopes, idiotope-paratope interactions result in dominant suppression for all antibody-producing cells. Entry of an epitope blocks Ab1 paratopes, resulting in the neutralization of their suppressive effect for the internal image set of Ab2 as well as of their stimulatory effect for the anti-idiotypic set of Ab2, which is suppressive for Ab1. Both effects thus favor the escape of Ab1-bearing cells from suppression. These assumptions allowed Jerne to propose reasonable qualitative interpretations of several standard reactions of the immune system, such as low zone tolerance, antigenic competition, 7-S inhibition, etc. Moreover, the network theory provided an

The idiotypic network theory

explanation for a rather disturbing observation recently published by Oudin and coworkers[16]: They had found that antibodies to two different epitopes of the same antigen sometimes shared idiotopes. According to the clonal selection theory antibodies to different epitopes, even if they were present on the same antigen, are the result of independently stimulated clones and bear no relation one to another. The clonal selection theory was thus unable to deal with Oudin's observation. In the network theory, the Ab1 sets potentially responding to the two epitopes, although they differ in paratopes, can be assumed to contain certain clones that share some idiotopes. As outlined above, neutralization of the stimulatory effect for the anti-idiotypic set of Ab2 in one of the two simultaneous Ab1 responses will result in escape from suppression of the Ab1 clones with shared idiotopes of the other response, and vice versa. Along the same line, the network theory postulated that antibody responses always contain, in addition to antigen-specific antibodies, a "parallel set" of antibodies that shares its idiotopes but does not react with the antigen. Induction of some non-specific antibodies had been observed in many experiments and the network theory offered an attractive and testable hypothesis for their generation.

The immediacy of the impact of the network theory on research in immunology was unprecedented. Only a small number of laboratories had been engaged in research on idiotypy before, as reflected in a low rate of 2–7 publications per year between 1968 and 1973. In 1974, when Jerne's first network paper appeared, the annual rate of papers dealing with idiotypy increased to 40, surpassing 100 in 1978, 200 in 1980, and 300 in 1982 (Fig. 9.3). Until today well over 5,000 publications mention the word idiotype or idiotypic in the title or abstract. The total citations of Jerne's 1974 paper alone amount to about 3,000 until today. Melvin Cohn, together with Peter Bretscher author of the competing associative recognition theory of 1970, was infuriated by the success of the network theory and vigorously attacked Jerne wherever he had the opportunity[17]: "I distinguish between a guess ("preconceived idea") and a theory (idea preconceived). As a minimum, the difference is that a theory is a Socratically argued guess. This is why the so-called … 'idiotype network' formulations are guesses not theories. Guesses may be right or wrong, provable or disprovable, brilliant or inept, clear or turbid. Unless they are Socratically argued in a way which ties them into a set of general principles, they remain psychological, not conceptual, contributions". Including the network collective in his critique, Cohn did not refrain from using ridicule: "What is the psychological reason that the scientific community expends such a disproportionate effort on the metaphysics (or illusion) of functional idiotype networks? A charismatic scientist who uses arguments of elegance and parsimony to convince us that a pure diet of chocolates cannot be fattening because they are hard to peel, can dominate thinking in a field where Lockean empiricism reigns; after all, the community argues, it must be true; we did the experiment and choco-

Figure 9.3
Numbers of publications per year containing the words "idiotype/idiotypic" and "antigen/antigenic" in the title or abstract.
The graph shows the dynamics of publications referring to "idiotype" (black triangles) compared to that referring to "antigen" (grey circles, × 100) between 1971 and 2005. Publications referring to "antigen" are more numerous because the term is more general than the term idiotype. The steady increase of publications referring to antigen until 1995 and its subsequent levelling off roughly reflect the overall dynamics of publications in the field of immunology. In contrast, the rapid increase and subsequent drop of publications referring to idiotype demonstrates the initial excitement and subsequent disappointment with a scientific fashion. (Data extracted from Thompson Scientific Web of Science).

lates are, in fact, hard to peel". In spite of these efforts, during the heydays of the network theory the impact of the associative recognition theory remained negligible, as evidenced by citation rates (Fig. 9.4). After its appearance in 1970 citation rates of the Bretscher/Cohn paper increased to about 40/year in 1974, thereafter dropping to below 20/year where they remained without exception until 1989. In contrast, citations of Jerne's first network paper[8] surpassed 50/year in 1979, 100/year in 1981, and 200/year in 1989. Only after 1990, the first year when a dramatic drop of citations heralded the downfall of the network theory, did the quotation rate of Bretscher's and Cohn's theory begin to increase from less than 20 to between 40 and 60/year (see also chapters 7, 17).

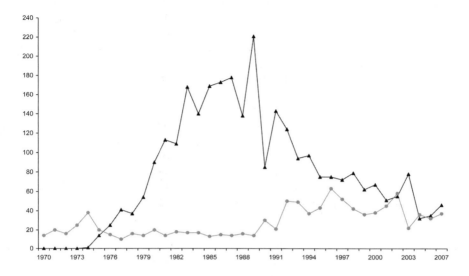

Figure 9.4
Frequencies of citations/year of Jerne's network theory compared to Bretscher's and Cohn's associative recognition theory.
The graph compares the dynamics of citations of Jerne's initial presentation of the network theory (black triangles, Jerne NK (1974) *Ann Immunol (Inst Pasteur)* 125C: 373) with that of Bretscher's and Cohn's initial presentation of the associative recognition theory (grey circles, Bretscher P, Cohn M (1970) *Science* 169: 1042) from 1970 to 2007. Note the initial increase in citations of Bretscher/Cohn until 1974, when Jerne's paper appeared, followed by a drop to near insignificance while Jerne's citations rose to impressive numbers, and the increase after 1990 when Jerne's citations began to fade (Data extracted from Thompson Scientific Web of Science).

The research stimulated by the network theory can be roughly divided into three phases. An early phase partially preceded and accompanied the development of the network theory, and was to some extent performed by researchers who closely interacted with Jerne, including members of the Basel Institute of Immunology of which Jerne was the director. Because there was a mutual influence between the early network research and the development of the theory, this phase will be included in this chapter. A second phase was characterized by a desire of a group of enthusiastic followers to confirm and expand the network theory, including the severe fallacies as far as the nature of the T cell receptor and of suppressor T cells are concerned (chapters 10, 11). A third phase was dominated by theoretical considerations which placed the network theory in a philosophical context, at the same time detaching it from is biological foundation (chapter 12). In this and the following chapters it will be unavoidable to discuss certain key experiments. With a few exceptions, they will be presented without much detail, emphasizing their cognitive contribution rather than their design. I will purposely refrain from any critique of the sort that is used in peer review, although this would certainly be appropriate in many cases. The purpose here is to analyze the development of the network collective as it went, not as it could have happened had there been a more critical attitude in the community.

The production of anti-idiotypic antibodies after reinjection of antibodies isolated from an animal back into the same animal had already been demonstrated[7]. The important question now was whether anti-idiotypic antibodies, i.e. antibodies specific for idiotopes of induced antigen-specific antibodies, were endogenously produced as a regular event in ongoing immune responses. The concomitant production of anti-idiotypic antibodies together with antigen-binding antibodies was positively demonstrated between 1974 and 1976 by three different laboratories in at least two experimental systems, one using a bacterial antigen (phosphorylcholine)[18,19] and the other using allogeneic MHC antigens[20] for immunization of mice. It should be noted, in this as well as in all other cases discussed later, that attempts yielding negative results were not reported. Experiments with negative results are usually regarded as meaningless, owing to technical or other insufficiencies. The number of variations in which an immunization experiment can be performed is infinitely large, and one can always argue that any other approach could have been successful. Moreover, and needless to mention, no scientific journal will accept a negative experiment for publication.

Testing predictions of the network theory in reproducible experiments required antibody responses with predictable idiotypes. Predictable idiotypes are by definition exceptional, since in most immune responses the produced antibodies differ between even genetically fully identical individuals, such as inbred mice. The idiotypes of the antibodies produced by immunization of an animal are therefore normally unpredictable, they are individually specific as the term idiotype implies. This represented a serious obstacle for network experiments, as the results obtained in one animal are usually not repeatable in a second animal. Luckily, as there are always exceptions to any rule in biology, network enthusiasts could make use of a number of experimental systems in which inbred mice were known to produce antibodies with similar idiotypes when immunized with certain antigens[21,22]. Most of these systems had been developed for the purpose of using idiotypes as genetic markers for antibody variable region genes, and there had been hot discussions whether or not idiotype sharing meant structural identity of the antibody variable regions (see chapter 8). In the network context this was not a critical point. Important was that each immunized mouse produced antibodies that reacted with the same anti-idiotypic antiserum, i.e. antibodies that shared at least some idiotopes. These experimental systems were collectively termed "crossreactive idiotypic systems"[23].

Antigens that induced antibodies with crossreactive idiotypes in inbred mice were either of bacterial origin (phosphorylcholine, dextran, streptococcal carbohydrate, staphylococcal nuclease, inulin) or small chemical antigens (haptens) coupled to protein carriers (arsonate, nitroiodophenyl). The reference antibodies against which the standard anti-idiotypic reagents were prepared were either isolated from the pooled sera of immunized mice, or were mouse myeloma proteins with the corresponding specificity. What made these immune responses so unique remained basically unknown. One property that seemed to play a role was the relative homogeneity of the induced antibodies, consistent with the assumption that the epitopes involved could interact with only a limited choice of paratopes.

Using crossreactive idiotypes the effects of anti-idiotypic antibodies (Ab2) on the generation of antibodies (Ab1) to a defined antigen could be studied in a reproducible fashion. Already in 1972 it was shown that injection of anti-idiotype antibodies into mice suppresses the production of antibodies with the target idiotype[10,11], suggesting that lymphocytes are susceptible to regulation by anti-idiotypic antibodies. In all his early network papers Jerne quoted these experiments as a key determinant in the development of the network theory. Similar results were later reported for several other systems, i.e. following injection of anti-idiotypic antibodies, the antibodies produced by subsequent immunization with the antigen failed to react with the anti-idiotypic antiserum[12,24]. The common interpretation of this result was that among the multiple clones potentially reactive to the antigen, those that expressed antibodies with the crossreactive idiotype were suppressed, while those that did not bear the crossreactive idiotype were free to respond and secrete their antibodies. Effective concentrations of injected anti-idiotypic antibodies were consistent with reasonable assumptions on similar effects of endogenous anti-idiotypic antibodies. This fitted nicely to the network theory which had postulated that the interaction with receptor idiotopes generates a suppressive signal for the antibody-producing cell.

While the suppression in adult animals was shortlived, that induced in newborn animals was maintained for life, reminiscent of neonatal tolerance and stimulating speculations on a role of the idiotypic network in self-tolerance[25,26]. Inactivation of arising clones by self-antigens, as proposed by the clonal selection theory, was anyway no longer a satisfactory explanation for the failure of the immune system to respond to self-antigens. Was self-tolerance perhaps established by idiotypic suppression in the network? Possibly, during the establishment of the network in neonates, self-reactive antibody receptors get integrated such that a permanent suppressive situation results for the clones bearing them.

Conversely, several laboratories reported experiments in which anti-idiotypic antibodies, depending on class or dose, induced an enhancement of the response to the antigen[24], or induced antibodies to the antigen even though no antigen was applied[27–29]. The observations on the stimulatory effects of anti-idiotypic antibodies were not readily reconciled with Jerne's original idea that idiotopes of antibody receptors were targets for suppression, unless one made the assumption that the stimulatory anti-idiotypic antibodies consisted predominantly of internal image set antibodies and stimulated idiotype production *via* receptor paratopes. Nevertheless, these results reinforced the suggestion that stimulatory and suppressive interactions coexisted in the idiotypic network.

In his second major paper on network theory[15], published in 1976 as the transcript of a Harvey lecture presented in 1975 at the Rockefeller University in New York, Jerne referred to the immune system as a "web of V-domains", presenting the basal concept of the immune system as an idiotypic network already as a fact of life. Here he introduced the Ab1, Ab2, Ab3 terminology (see above) and, while he did not add much in terms of mechanistic precision, he emphasized the cybernetic context by quoting extensively from several recent articles on general mathematical models on the function of dynamic multispecies communities[30,31]. Such models had been derived for ecological communities that consist of multiple species interacting with one another in either positive or negative ways. For example, in a wildlife community predator species feed on prey species, which is positive for the former and negative for the latter. From these models he adapted the cybernetic term "connectance" which denotes the proportion of species in a multispecies community with which one member species can interact. The "stability", also a cybernetic term, of a multispecies system depends on its connectance in such a way that a certain intermediate connectance generates stability, whereas values below or above result in instability. Interactions can vary from strong to feeble and, as far as stability is concerned, a few strong interactions can be equivalent to many feeble interactions. Jerne suggested that models on dynamic multispecies communities "may well have some relevance also for the network of the immune system".

Indeed, one did not have to wait very long until the first mathematical formulations of the network theory were put forward. Already in 1975 two such mathematical models appeared in press, independently developed by two young theoreticians, P.H. Richter and G.W. Hoffmann, who closely interacted with Jerne. Richter's model[32] was based on the assumption that the immune system was potentially connected by idiotypic interactions of V-domains. In the absence of antigen, or as long as certain assumed threshold concentrations of antigen are not reached, the system is at rest. Antigen concentrations above threshold stimulate Ab1, Ab2, Ab3, etc., in decreasing concentrations until subthreshold concentrations are reached. Reversely, Ab3 suppresses Ab2, which suppresses Ab1. Richter's differential equations described three dynamic states: Low zone tolerance, immune response and memory, and high zone tolerance. Low zone tolerance occurs when the concentration of antigen is sufficient to drive the response until Ab2 but no further. As a result, Ab2 suppresses Ab1 but is itself not subject to suppression, so that Ab2 dominates over Ab1. An immune response takes place when the chain reaction proceeds until Ab3, which suppresses Ab2, allowing Ab1 to predominate so that Ab1-producing cells can proliferate and develop into memory cells. High zone tolerance is generated when the chain reaction proceeds until Ab4, or further, so that suppression results for Ab1. The Richter model, although it was mathematically correct and explained several standard reactions of the immune system, was intuitively judged by most experimental immunologists as too reductionist and academic, as it ignored a number of established facts such as, for example, the T cell/B cell dichotomy. It was never widely accepted as reflecting reality or worth testing experimentally.

More elaborate was Hoffmann's model[33], which also accepted the idiotypic network as a basal fact, but acknowledged the existence of T and B cells and incorporated a greater number of experimental observations, including recent findings that T cells can produce antigen-specific soluble factors which regulate the immune response[34]. The model radically simplified the idiotope-paratope concept of V-domains by defining only two complementarities with respect to an antigen, a "positive set" of receptors which is complementary to the antigen, and a "negative set" of receptors which is complementary to the positive set. The positive and negative sets are complementary and bind to one another. B cells express bivalent antibodies as receptors whereas T cells express monovalent antibody-like receptors. T cells are readily activated by antigen to proliferate and to secrete their monovalent receptors as factors which can then bind to the antigen or to the opposite receptor sets on T and B cells. B cells are less readily activated and proliferate less upon stimulation by antigen. They secrete divalent antibodies which bind the antigen as well as the opposite receptor sets on T and B cells. Central assumptions of the model are that T and B cells are potent stimulators of the opposite sets owing to the high receptor density on their surface. In contrast, binding of a monovalent T cell factor to a receptor results in neutralization by blockade, whereas binding of a divalent antibody to a receptor results in cell killing. The model formulates a set of differential equations which define a number of steady states, including the virgin state, immunity, memory, tolerance, etc. In contrast to Richter, Hoffmann developed his model further over many years until to date. It was refined by incorporating most novel developments in experimental immunology, and adapted to model pathological situations such as autoimmunities and immunodeficiencies, including AIDS. Hoffman remains convinced that his model accurately describes the functioning of the immune system.

A third mathematical model based on network theory was formulated by G. Adam and E. Weiler[35], with the aim to model the generation of B cell clonal diversity during ontogeny rather than the reactions of the immune system to antigens. Weiler and coworkers had made the puzzling observation that lymphocytes transferred into a syngeneic mouse would fail to react to an antigen unless the recipient mouse was irradiated. This phenomenon was termed "permissiveness" and interpreted such that the intact immune system of one individual mouse was non-permissive for the lymphocytes of another mouse. The main assumption of Adam and Weiler was that permissiveness required idiotypic compatibility which was established during early ontogeny among random somatic mutants of dividing lymphocyte clones. Making the assumption that clones proliferated more rapidly than mutations could be accumulated, Adam and Weiler calculated that the actual clonal diversity generated in an individual mouse was two to three orders of magnitude smaller than the potential clonal diversity. Each mouse thus developed its own individual clonal repertoire as an idiotypically compatible network. Addition of cells belonging to another network would result in idiotypic incompatibility and functional incompetence.

Jerne, who savored speculations and imagination in science, happily discussed these models in his Harvey lecture, which he closed with the remark: "Viewing the immune system as a network has provided a fertile basis for speculations and, it is hoped, will lead to a better understanding of the rules by which the immune system develops, functions, and is regulated".

Chapter 10
The T cell receptor puzzle

...antibodies were ... the main, if not the only element on which all immunological research was focussed. Phenomena which did not involve antibodies were thought not to belong to immunology.

Niels K. Jerne (1989)
Progress in Immunology VII.
Springer Verlag, Berlin

As long as this was possible, Jerne was a firm believer in the notion that antibodies are the sole molecular basis of immunological specificity and of idiotypic network interactions[1,2]. In defending this notion, he had to deal with recalcitrant implications derived from the genetic control of immune responsiveness, notably the growing evidence that essentially all Ir genes were found to map to the MHC. Since Ir genes controlled the responsiveness of T cells in an antigen-specific fashion, what was more plausible than to propose that the antigen receptors of T cells were the direct products of these Ir genes, and thus of genes linked to the MHC? These ideas were put forward by prominent immunologists including Hugh McDevitt and Baruj Benacerraf, both pioneers in studies of Ir genes[3]. Since antibody genes were definitely not linked to the MHC, notably Benacerraf and coworkers suggested that T cell antigen receptors, on cells and in secreted form, were different from immunoglobulins and represented another, second class of antigen-specific receptors in the immune system[3–5]. A fundamental difference between the receptors of T cells and antibodies was further supported by experiments showing that T cells recognized protein antigens differently from antibodies and B cells. Whereas B cells reacted to conformational epitopes that were destroyed by denaturing the protein, T cells reacted to sequential epitopes that were resistant to denaturation.

Jerne opposed these suggestions vehemently in all his papers and lectures. Here are some quotes: "Benacerraf and McDevitt regarded inescapable the conclusion that there exists a class of molecules encoded by these Ir genes, which are responsible for the recognition of specificity at the T-cell level, and that these molecules are not immunoglobulins. Benacerraf and McDevitt called their proposal revolutionary and heretical, respectively, and I suggest that this proposal should be resisted as long as other ways of interpretation remain possible, that do not sacrifice the idea that immunological specificity resides solely in antibody molecules"[2]. Another quote: "Some research workers claim to have evidence indicating that T cells produce a novel class of antigen-recognizing molecules, (a) which differ in structure from antibody molecules, (b) which may be univalent and presumably do not possess the V-domains that form paratopes ... of classical antibody molecules, and (c) which recognize a universe of epitopes ... that differs from the universe of epitopes recognized by antibodies. I believe we should remain skeptical of these notions until this class of T-cell receptors has actually been isolated and characterized. It seems likely to me that the continued search for T-cell receptors will reveal that they present at least the V-domains of classical antibody molecules"[1].

As an alternative explanation for the role of MHC-linked Ir genes in the control of immune responsiveness Jerne offered the mechanism that he had proposed in his theory on the somatic generation of immune recognition of 1971[6] (see chapter 7). In short: Inherited ("germline") antibody genes encode antibodies with specificity for all MHC molecules (including the Ir-gene products) of the species; the majority of B cell clones produce antibodies against the MHC molecules not present in the individual, can develop unperturbed, and are responsible for the strong reactivity against allogeneic MHC antigens in transplantation; B cell clones that produce antibodies against the individual's own MHC molecules are self-reactive and are forced to mutate their antibody genes in order to lose self-reactivity and ensure tolerance of self-MHC molecules; these somatically mutated antibody genes encode the repertoire of antibodies that recognize foreign antigens. By determining which of the inherited antibody genes have to mutate, the MHC thus indirectly shapes the repertoire of antibodies available for reactions to foreign antigens. In order to explain Ir gene control, "all that would be needed is to postulate that the acquisition of tolerance to an Ir-gene product leads to T cell responsiveness to a set of antigens"[6].

With respect to the nature of the T cell receptor, neither Benacerraf and McDevitt nor Jerne were correct, as will be discussed below. However, the general fascination with the network theory stimulated a wave of research aimed at proving Jerne's claim correct, namely that immunological specificity was mediated solely by the V regions of antibodies. It should be noted here that the network theory did not absolutely require antibodies as sole carriers of immunological specificity. A functional network could as well be envisaged to comprise two or more classes of antigen-specific receptors in the immune system, as long as they possessed variable regions that exhibited idiotopes and paratopes and were able to react with one another by

The T cell receptor puzzle 97

paratope-idiotope interactions. As a result, the eventual elucidation of the non-immunoglobulin nature of the T cell receptor did not kill the network theory. Nevertheless, an immune system only based on antibodies was more puristic and esthetic, and Jerne's strong statements in this direction seemed to have induced a desire in many immunologists to see the immune system in Jerne's way.

Most of these experiments, in one way or another, were designed to demonstrate that antisera prepared against immunoglobulin molecules would react with T cell receptors. The situation was difficult as it was clear that anti-immunoglobulin antisera which gave a strong reaction with B cells, easily visible for example in a fluorescence microscope, failed to show similarly unequivocal reactions with T cells[7]. Indeed, fluorescence microscopy of T cells with anti-immunoglobulin antisera gave variable results such that some antisera reacted positively whereas others failed to react. Thus, if T cells used antibodies as receptors, they did not display them on their surface in a readily detectable way as did B cells. For example, it was possible that the receptor antibodies on T cells did not stick out as far as they did on B cells, so that they were not accessible to all anti-immunoglobulin antibodies. Experimental designs thus had to be refined to more sensitive methods of detection. One such method was the use of anti-immunoglobulin antisera to block antigen recognition by T cells, and several groups reported positive results of such experiments[8,9]. Another method was based on the observation that antibody receptors of B cells could be made to redistribute by exposure to anti-immunoglobulin antibodies[8]. They changed from an even distribution over the entire cellular surface to form a single cap on one pole of the cell which gave very bright fluorescence due to the high reagent density in the cap. Also by this method, the detection of immunoglobulin molecules on T cells was reported by several investigators.

Another technology, very advanced at the time, for demonstration of immunoglobulin-like molecules on T cells was the immunoprecipitation of surface-radiolabelled cells with anti-immunoglobulin antisera followed by separation in electrophoretic slabs. In addition to the serological reactivity with anti-immunoglobulin antibodies, this technique added the molecular weight, estimated by electrophoretic mobility, as a biochemical parameter to the analysis. A number of groups reported to have recovered immunoglobulin-like molecules from T cells or T lymphoma cells, with some but not all anti-immunoglobulin antisera. Most of these results, obtained for several species including man, mouse, rabbit, birds, and fish, were consistent with the conclusion that T cell receptors consisted of antibody-like molecules with a novel class of heavy chain that was related but perhaps not identical to that of IgM[10].

The network theory only made sense if all specific elements of the immune system, including antibodies, B cells, and T cells, were connected by idiotypic interactions. Accordingly, experiments were performed to demonstrate such interactions, which could be designed in three different versions:

(i) To reveal reactions of antibodies against antibody idiotypes with receptors of T cells, (ii) to reveal reactions of antibodies against idiotypes of T cell receptors with B cell idiotypes, and (iii) to reveal reactions of anti-idiotypic T cells with B cell idiotypes. The first of these versions was also the first to be reported as successful, as it could be performed with anti-idiotypic antisera already available from studies of crossreactive idiotypic systems. For example, it was reported that injection of such antisera into mice of the strain that produced the appropriate crossreactive idiotype activated helper T cells in these mice[11]. The helper T cells recognized the same antigen as the antibodies that carried the crossreactive idiotype and could be demonstrated in adoptive transfer experiments in which the antigen was given as carrier to which a hapten was coupled for recognition by B cells. This type of experiment was performed in several crossreactive idiotypic systems and suggested that the injected anti-idiotypic antibodies reacted with idiotopes of the receptors of T cells and activated them. The experiment thus demonstrated the existence of an idiotypic connection between antibodies and the receptors of T cells. Moreover, the fact that at least some of the activated T cells recognized the same antigen as the antibodies was a strong indication for the structural relatedness of the T cell receptors to antibodies, i.e. not only did they share the idiotopes but also the paratopes of the crossreactive idiotypic system[12].

While the initial attempts to relate T cell receptors to antibodies relied heavily on serology, it was soon felt that additional parameters would strengthen the significance of the results obtained. What followed were experiments in which serology was combined with Mendelian genetics, in which it was shown that T cell receptor idiotypes were controlled by genes linked to the antibody heavy chain genes. The knowledge about the genes encoding the antibody heavy chains at the time stemmed largely from studies of immunoglobulin allotypes and was fragmentary and largely incorrect. It was thought that the "heavy chain linkage group" contained two separate but linked clusters of many V genes and a few C genes, respectively. No other genes were known in the cluster, so that Mendelian linkage of a phenotypic character to the heavy chain genes was taken as a strong indication that the phenotype was encoded by a heavy chain gene. In case of idiotypes, which are located in the variable region of antibody polypeptide chains, this meant that they were encoded by antibody V genes. Experiments of this type in several crossreactive idiotypic systems showed that T cell receptor idiotypes were under genetic control of antibody heavy chain genes, giving rise to the interpretation that T cell receptors were encoded, in part, by antibody VH genes[13].

In a particularly influential series of experiments receptor molecules were isolated from mouse or rabbit T cells by adsorption of the cells to antigen-coated nylon fibers and subsequent desorption by temperature shift[12,14]. Receptor molecules remained attached to the fibers and could be recovered by mild denaturating conditions. The presence of functionally

The T cell receptor puzzle

active T cell receptors in these preparations was demonstrated by inhibition of the infectivity of bacteriophage to which the antigen – in this case various haptens such as nitroiodophenyl – was coupled. Using this exceedingly sensitive assay, the miniscule amounts of recovered receptors were then characterized for affinity, specificity, serological markers including allotypes and idiotypes, etc. The main results of these experiments were that T cell receptors carried serological, genetic, and functional markers associated with the heavy chain V regions of antibodies of the same specificity, but no other markers of immunoglobulins[15,16].

Another highly influential series of experiments utilized allogeneic antisera produced against the idiotypes of rat T cells reactive with allogeneic MHC antigens[17]. These antisera against allo-MHC-reactive T cell receptors cross-reacted with allo-MHC-reactive B cell receptors, an example demonstrating the similarity of T cell receptor idiotopes with that of B cell receptors (see above, version ii). Soluble receptor material reactive with the anti-T cell receptor idiotype antisera could be found in normal rat serum and urine. The expression of idiotype positive material was genetically polymorphic and its inheritance was controlled by genes linked to the antibody heavy chain genes. The authors were happy to report that "These data on genetic linkage to the Ig heavy chain genes are in complete accord with data obtained in the mouse on the inheritance of helper T cells using heterologous anti-idiotypic antisera"[18].

Altogether the results from several unrelated experimental systems seemed to agree with one another. This made a strong impact and was instrumental in the then widely accepted notion that T cell receptors, while being different from antibodies, contained the variable region of antibody heavy chains as a shared structure determining specificity and idiotype. Of note, this notion strikingly confirmed Jerne's minimal prediction on the nature of the T cell receptor (see above). Even scientists who otherwise could only be convinced by structural data, such as Gerald Edelman, joined in the chorus. In his Summary of the 1976 Cold Spring Harbor Symposium on Immunology, Edelman concluded: "It appears that the T cell receptor is an immunoglobulin, at least in the sense that it has V regions at its active site. We are thus relieved of the repellent possibility of having to invoke two separate origins of diversity, one for B cells and one for T cells"[19].

What were the reasons that made so many highly intelligent scientists commit the same errors and draw the same false conclusions? Was it perhaps the wish to fulfill Jerne's prophecy? Many of the experiments demonstrating idiotypic crossreactivity between antibodies and T cell receptors merely by serology were done either *in vivo* or with super-sensitive assays designed for detecting minimal quantities. The evidence was thus extremely indirect and corrupted by an enormous black box. Their theory-ladenness not only permitted but indeed demanded alternative interpretations, but alternative interpretations were rarely taken into consideration. Already at the time it was clear that serological crossreactivity was far from proof of structural identity. The interpretation in terms of structural similarity/identity of T cell receptors with antibodies or parts thereof was thus heavily influenced by the investigator's wish to obtain the desired results. Nevertheless, with hindsight, it is not to be excluded that many of these results had been correct and that only the interpretations had been erroneous. For example, serological crossreactions, as detectable by antisera, can involve antibodies to polysaccharide side chains or to short conformational determinants that can readily be envisaged to exist and to be similar between T cell receptors and antibodies.

As far as the results showing genetic control of T cell idiotopes by VH genes are concerned, it is more difficult but also possible to argue that they had been correct and only erroneously interpreted. When in 1984 unequivocal evidence for an unrelated set of T cell receptor genes came forward[20,21], the VH linkage of T cell receptor idiotypes could still be rescued because the genetic control of a phenotypic character by a certain gene or group of genes does not necessarily mean that the character is directly encoded by the same gene(s). The argument was borrowed from the Adam-Weiler Model[22] of the network theory, in which the immunoglobulin variable regions serve as selective elements in the development of the B cell receptor repertoire. Extending this role to the T cell receptor repertoire would result in the selection of T cell receptors by interaction with immunoglobulin V regions. This selective influence could result in the phenotypic linkage of T cell receptor idiotypes to the heavy chain allotype, without any genetic linkage of the genes directly coding for these proteins. Indeed, this argument has later been made to defend the results showing control of T cell receptor idiotypes by VH genes[23], and experimental evidence in its support has been produced even recently[24].

The phenomenology around the T cell receptor in the second half of the 1970s indeed presented immunologists with an unsurmountable intellectual challenge. The VH gene control of T cell receptor idiotypes had become widely accepted knowledge, as was the control of T cell antigen-specific responses by MHC-linked Ir genes. Moreover, Rolf Zinkernagel and Peter Doherty had just discovered the phenomenon of MHC-restricted antigen recognition of T cells[25]. They had found that virus-specific cytotoxic T cells that arise in mice infected with Lymphochoriomeningitis Virus (LCMV) lyse LCMV-infected target cells only if they express the same MHC antigens as the infected mouse. Analogous results were not much later obtained for helper T cells (see chapter 7)[26–28]. This meant that T cells had two specificities, one for the antigen and the other for syngeneic MHC gene products. Nobody understood how these two specificities came about and so, as usual in immunology, alternative models were proposed and the scientific community split into factions vehemently defending one or the other model. One faction postulated two types of receptors per cell, the other favored one receptor recognizing a combined MHC-antigen complex referred to as "altered self". An additional line of evidence concerned soluble factors,

The T cell receptor puzzle

derived from T cells and replacing their functions as suppressor cells or helper cells in appropriate assays[5]. These factors had antigen binding specificity and reacted with antisera raised by immunization between MHC-disparate mouse strains (see next chapter). These factors thus seemed to represent secreted T cell receptors. Network enthusiasts could not ignore these findings which were published in the proper scientific journals by respected colleagues.

Retrospectively it is likely that many of the results on T cell receptors of the time had involved errors not only of interpretation but of experimental data processing as well. In this context I will discuss one of my own published experiments which may serve as an example how the wishful choice between different ways to look at the same data can bias the result in favor of a desired outcome. As an exception, this experiment will be presented in somewhat more detail. To account for the influence of the two types of genes (MHC and VH) on T cell specificity, it was hypothesized that T cell antigen recognition involved two types of polypeptides, one partially encoded by VH genes and the other by MHC-linked genes. We reasoned that with anti-idiotypic antibodies prepared against T cell receptors the combined influence of the VH genes and of the MHC should somehow become apparent. We even hoped that the results would allow us to distinguish between one- and two-receptor models of T cell antigen-recognition. To this end, my colleague Peter Krammer and I decided to study the idiotypes of allo-MHC-reactive T cells generated in AKR mice against the MHC antigens of strain B6. An anti-idiotypic antiserum was produced against these AKR T cells in (AKR×B6) F1 mice, which are devoid of AKR anti-B6 T cells and therefore able to generate antibodies to their receptors. The anti-idiotypic antiserum was found to react, in immunofluorescence, with 35% of activated AKR anti-B6 T cells but only 5% of the anti-B6 T cells of strain SJL which differs from AKR in both MHC (the H-2 locus) and heavy chain allotype (the Ig-1 locus). The 5% was taken as "background" and the idiotype was declared as specific for AKR anti-B6 T cells. The genetic segregation and linkage of the idiotype was then tested in a backcross experiment in which (AKR×SJL) F1 mice were crossed with SJL mice. Because the H-2 and Ig-1 loci segregate independently of each other, the backcross offspring segregated into four groups of about 25% each, group 1 homozygous for SJL alleles (H-2s, Ig-1b) at both loci, group 2 with one Ig-1d allele of AKR, group 3 with one H-2k allele AKR, and group 4 with both one H-2k and one Ig-1d allele of AKR (see Tab. 1). T cells of each group were pooled and stimulated with B6 cells in mixed lymphocyte cultures. When tested for reactivity with the anti-idiotypic serum, 40% of the T cells of group 4 – that inherited both the Ig-1d and the H-2k alleles of AKR – but only about 6% of each of the three other groups were positive. We declared the 6% as "background", and so the data unequivocally suggested that the inheritance of the T cell receptor idiotype required both the heavy chain as well as the MHC genes of AKR. This was one of two alter-

102   Chapter 10

Table 1.
Linkage of T cell receptor idiotype to the *H-2* and the immunoglobulin-1 locus

T cell blasts from MLR* Responder	Stimulator	Positive responder T cell blasts with Fla (AKRaB6) anti-idiotypic serum (%)
Group 1 (H-2$^{s/s}$, Ig-1$^{b/b}$)	B6	6.2%
Group 2 (H-2$^{s/s}$, Ig-1$^{b/d}$)	B6	6.6%
Group 3 (H-2$^{s/k}$, Ig-1$^{b/b}$)	B6	6.3%
Group 4 (H-2$^{s/k}$, Ig-1$^{b/d}$)	B6	40.6%

The experimental conditions for immunofluorescence are the same as in Table 1. All fluorescent slides were read blind. Control staining with normal (AKR×B6) F1 serum: 1.7% (group 1), 0.6% (group 2), 0.6% (group 3), and 4.6% (group 4).
* Responder cells were nylon wool column passaged nonadherent lymph node cells (over 99% immunoglobulin negative by immunofluorescence; 40% (group 1), 64% (group 2), 48% (group 3), and 56% (group 4) recovery of cells from the nylon wool column. Group 1 (R), B6 (S): S. J. 143; group2 (R), B6 (S): S.J. 319; group 3 (R), B6 (S): S.J. 64: group 4 (R), B6 (S): S.J. 128.
** Each group consisted of cells from a pool of 4 mice H-2 typed and allotyped as described 8,9.
From Krammer PH, Eichmann K (1977) T cell receptor idiotypes are controlled by genes in the heavy chain linkage group and the major histocompatibility complex. *Nature* 270: 733–735. Table 2

native anticipated results and the paper was written in support of a one-receptor model of the T cell receptor.

What was not considered in analyzing the data was the possibility that the 40% vs. 6% positive cells in the backcross mice reflected the overall proportions of B6-specific T cells in the 4 mixed lymphocyte reactions, a parameter that remained unknown because it was not possible to determine at the time. If that had been the case, the formally correct way of analyzing the data would have been to present them as ratios of % idiotype-positive cells over % of cells of each group reactive with normal control serum (these values are listed in the footnote of Table 1). Had we chosen this way of analysis, we would have obtained the following results:

Group 1: 3.6; Group 2: 11.0; Group 3: 10.5; Group 4: 8.8.

These results would have meant that no significant difference existed between groups 2, 3 and 4, while group 1 had a somewhat lower value but of questionable significance with respect to the difference to the other groups. The data would thus have been inconclusive, and in retrospect are compatible with any interpretation, including the possibility that the antiserum reacted with some T cell activation-antigen rather than an idiotype. Had we considered this possibility, these results would probably not have been published, certainly not in a journal like *Nature*. Instead we were so flabbergasted by getting the desired result that a critical consideration of other possibilities, of which there were still more than the example described above, was not even attempted. In his book on the generation of scientific facts Ludwik Fleck described this phenomenon of selective perception as Gestalt-vision, using the historical variations in the drawings of human anatomy as one of his examples, and concluded: "Observation without assumption, which psychologically is nonsense and logically a game, can be dismissed".

Another example of Gestalt-vision is this: Both in the areas of Ir gene control of T cell reactivity and VH linkage of T cell idiotype, selective influences of the MHC and/or the heavy chain V regions on the T cell repertoire were initially rarely considered, although this would have made interpretations a lot easier. McDevitt's and Benacerraf's suggestion that the T cell receptor was encoded by MHC-linked Ir genes was as wrong as was the notion of heavy chain allotype linkage of the T cell receptor genes. The declared purpose of this research was to determine the molecular identity of the T cell receptor, and so data were interpreted with respect to genes and protein structure. Alternatives such as selection in ontogeny would have served equally well to interpret the data, but were not considered because they did not speak to the aim of the experiments. Selection was not taken into consideration until Zinkernagel and colleagues found that MHC restriction did not require the MHC allele to be expressed in the cytotoxic cells that expressed the relevant T cell receptors. Rather, it was the MHC allele of the thymus in which the cytotoxic cells had matured that had to match with that of the target cell[30]. This and similar experiments uncovered the fact that the T cells, while maturing in the thymus, receive a survival signal if their T cell receptor interacts with a low affinity with the MHC antigens expressed in the thymus. If there is no interaction or if the affinity is too high, the T cells receive no signal or too strong a signal and do not survive. As a result, mature T cells that leave the thymus are selected for low affinity interaction with self-MHC molecules.

The biological significance of this selection process for the understanding of Ir gene control of immune responses remained obscure, until evidence from several independent lines of research, structural, biochemical, and genetic, came together and made things fall into place. The first was the elucidation of the 3-dimensional structure of MHC molecules by X-ray crystallography[31,32]. This work uncovered the fact that MHC molecules possess a pocket in which they carry small peptides which they pick up during their biosynthesis inside the cell. After travel to the cell membrane, the MHC/peptide complexes are displayed on the cell surface. While most of the peptides stem from the physiological turnover of endogenous proteins, foreign protein antigens that enter a cell by infection or phagocytosis are also degraded, resulting in the presentation of foreign peptides by self-MHC molecules on the cell surface. MHC molecules in the thymus present mostly endogenous peptides so that T cells are selected for low affinity interaction with self-MHC molecules presenting endogenous peptides. By way of probability, however, a minute proportion of them express receptors that interact with high affinity if the MHC molecule presents a given foreign peptide. While the low affinity interaction is insufficient to activate the T cell, the high affinity interaction leads to the activation of T cells which then proliferate and differentiate into functional helper or cytotoxic T cells, depending on the type of MHC molecule on which they have been selected. These results suggested an alternative interpretation for the influence of MHC-linked Ir genes on immune responsiveness: Some pockets of certain MHC alleles are unsuitable to present certain peptides, resulting in nonresponsiveness to those peptides of the helper T cells or cytotoxic T cells selected on these MHC alleles. Rather than Ir genes encoding T cell receptors, this explanation proved to be correct.

Of pivotal importance in the molecular identification of the T cell receptor was the availability of the technique for producing monoclonal antibodies, invented by Georges Köhler and Cesar Milstein in Cambridge, UK, in 1975[33]. By the end of the decade this technique had been adopted by many laboratories and monoclonal antibodies had replaced the much more error-prone conventional antisera that had been used before. While all of the misleading results on T cell receptor idiotypes had been obtained with conventional antisera, several laboratories began to produce monoclonal antibodies by immunization of mice or rats with mouse T cells, and fusing the B cells of the immunized mice with appropriate tumor cell lines to generate hybridomas. As in many other applications of serology, monoclonal antibodies brought the breakthrough in T cell receptor research as well[34-36]. In 1983 several laboratories had generated monoclonal antibodies which identified two different components of the T cell receptor, the invariant CD3[36] molecule and the variable β chain[34,35]. Biochemical studies using these antibodies identified a complex in the T cell membrane that consisted of several small invariant polypeptides, the CD3 complex, and a heterodimer of two polypeptide chains, α and β, which both displayed signs of molecular diversity and were thus strong candidates for antigen recognition.

Adopting the hybridoma technology for the generation of monoclonal antigen-specific T cell lines provided the material to search for and identify the T cell receptor genes. As immunoglobulin gene rearrangement had been described and methods were available to identify immunoglobulin heavy and light chain gene segments and their mRNAs, it did not take long until it was unequivocally shown that antigen-specific T cell lines did not rearrange or express immunoglobulin genes[37]. Using the technique of subtractive cDNA hybridization, which allows to identify genes expressed in a particular cell type but not in other cells, a number of T cell-specific cDNA clones were found that hybridized to a region in genomic DNA that has rearranged in the T cell line[20,21]. Such cDNA clones displayed V, D, J, and C regions analogous to immunoglobulin heavy chain cDNAs, but were otherwise structurally different. These initially identified cDNAs were found to encode the β chain of the $\alpha\beta$ heterodimer reactive with monoclonal antibodies to the T cell receptor. In due course cDNAs encoding the a chain, homologous to the immunoglobulin light chain, as well as the genes encoding a second class of heterodimer, the γδ T cell receptor, were found. The genomic organization of the T cell receptor loci was elucidated, showing the same principle of rearranging gene segments as known for the immunoglobulin loci. These loci were on chromosomes different from that encoding the immunoglobulins, and none of their segments, including any of the V genes, was shared between T and B cells. The notion that T cell receptors shared with antibodies the heavy chain V region was blown. What Edelman had referred to as "the repellent possibility of having to invoke two separate origins of diversity"[19], had indeed materialized, like it or not.

Chapter 11
Suppression turned idiotypic

I have become increasingly convinced that the essence of the immune system is the repression of its lymphocytes.
N. K. Jerne (1974) Towards a network theory of the immune system.
Ann. Immunol. Inst. Pasteur, 125 C: 373

Suppressive effects of lymphocytes in immune responses had first been observed in the early 1970s (see chapter 7). The phenomenon of cellular immune suppression thus preceded the network theory and research on suppressor cells was carried out initially independent of network concepts, as a paradigm in its own right. In these initial experiments suppression was mostly associated with T cells from various lymphoid organs that, admixed to antigen-stimulated lymphocyte populations, diminished their reactivity as determined by T cell proliferation assays[1–3]. Thus, in these initial experiments both the suppressive and the antigen-reactive lymphocytes had been T cells, so that suppression was perceived as an affair among T cells only. However, the notable exception among these early observations was a report on allotype-specific suppression of immunoglobulin production by suppressor T cells, suggesting that B cells, even without antigen-stimulation, can be a target of T cell suppression as well[4]. The idea that in order to produce antibodies a B lymphocyte first has to escape from the suppressive effect of suppressor T cells fascinated Jerne, who speculated that: "T cells recognizing the idiotypes of B cell receptors may be assumed likewise to maintain B cell suppression. Conversely, we could conclude that, normally, B cells remain functional because of the absence of sufficient numbers of specific suppressor T cells..."[5] Jerne's speculation provided the incentive to merge the two paradigms, suppressor T cells and idiotypic network theory. The ensuing rush of experimental activities eventually accumulated so many false results that, together with the errors in T cell receptor research described in the previous chapter, it became a major factor in the eventual rejection of the network theory as non-productive and superfluous.

Suppression turned idiotypic

Nowhere has immunological research been caught in so many traps as in the attempts to unravel the nature and function of suppressor T cells. In this chapter I will discuss the main areas of fault in this field, which included the excessive subsetting of suppressor T cells, the largely imaginary soluble "suppressor factors", and the I-J region of the mouse MHC, all subsequently proven to be nonexistent. Together with idiotype expression, VH gene usage, and VH-restriction of suppressor T cells and their factors, these studies amounted to what now appears as one of the major fallacies of modern life science.

Although cellular immune suppression had been initially discovered in several laboratories, it was Richard "Dick" Gershon who advanced the field more than anybody else and no other area of study in immunology – with the exception of Jerne's network theory – has been so radically personalized as T cell suppression with Gershon. Intriguingly, both of these highly personalized paradigms suffered, after a period of blossom, unprecedented downfalls, much more severe and not comparable with the well known paradigm shifts that occur as part of normal scientific progress.

In the 1970s the subdivision of lymphocytes into subpopulations developed as an important concept in immunology, i.e. functionally different types of lymphocytes could be distinguished on the basis of differences in the expression of surface markers. The field was kicked off by the discovery in 1970 (see chapter 7) that T cells expressed Thy-1 whereas B cells expressed immunoglobulin on their surfaces[6]. While these two fundamentally different types of lymphocytes looked exactly alike in a light microscope, the expression of different molecules on their surfaces now permitted their distinction by immunofluorescence microscopy or complement-dependent cell killing procedures using appropriate antibodies. Important to realize, for more than a decade most of lymphocyte typing was done by conventional antisera[7,8] and it was not before the late 1970s that monoclonal antibodies and flow cytometry slowly began to enter the field, eventually turning it into a reliable methodology[9,10]. This, however, was a slow process and until well into the 1980s the use of conventional antisera and subjective assays was common practice. Nevertheless, it should be stressed that the antisera in use in this field were mostly produced by immunization between genetically different mouse strains, restricting the formation of antibodies to the few antigens that were polymorphic in mice. In the appropriate strain combinations these antisera could be reproducibly generated in any experienced laboratory and, when used in the proper context and with care, they were reliable reagents. This is most impressively documented by the fact that many of the T cell surface antigens identified by the so-called allo-antisera can now be found among the CD cluster antigens defined by monoclonal antibodies.

An important set of polymorphic cell surface antigens, allowing the subdivision of T cells into helper and killer T cells, was the Ly series: Whereas helper T cells expressed Ly1, killer T cells expressed Ly2 and Ly3[11,12] (the latter corresponding to the CD8α and CD8β antigens, respectively, later defined by monoclonal antibodies). Several other series were in use as well, termed Qa, TL, and the like, together permitting to define a surface marker profile, the so-called phenotype, of a subpopulation of lymphocytes. Pioneering in this field was a cooperation between Edward "Ted" Boyse who knew how to make antisera and Harvey Cantor who had experience in T cell assays[13]. The notion that emerged from these studies was that each immune function is carried out by a specialized subset of lymphocytes, which can be identified by its surface phenotype. The more markers were to be found, the finer the subsetting and the better the phenotype-function correlation would become. It soon became clear that the subsetting of the T cell population would not end with the phenotypic distinction of helper and killer T cells. In this thought style it was obvious that suppressor T cells needed to have their own phenotype, distinct from that of helper and killer T cells[14].

Before monoclonal antibodies and preparative flow cytometry became common practice, experiments to define the surface phenotype of a functionally defined lymphocyte subpopulation were done mostly by treatment of spleen or lymph node cells with an antiserum and complement. The cells expressing the surface marker in question were thus depleted and the remaining cells were added to appropriate assays in which the function in question could be tested, such as cell culture systems or adoptive transfer into syngeneic mice. In a hypothetical experiment, lymph node T cells from a mouse that had previously been immunized with antigen X would proliferate in culture when antigen X was added; this proliferation could be inhibited when spleen cells from a normal mouse were added; when the spleen cells were pretreated with anti-Ly2 antiserum and complement, their suppressive effect was abolished; consequently, the suppressor T cells present in the normal spleen expressed Ly2 as a surface marker. It should be realized by the reader that such experiments, straightforward as they may seem, are associated with an enormous black box. This begins with the very labile activity of the complement in the rabbit sera that were mostly used as a source of complement. Neither complement activity nor the success of complement-dependent depletion could reliably be assessed because the appropriate assays were not available or too cumbersome to be included in every experiment. Further, the remaining cells in depleted cell populations consisted of a Gemisch of multiple other cell types including non-lymphoid cells, largely unknown in composition and function, not to speak of individual variations among donor animals, and so on. Consequently, experiments which failed to yield the expected result were often disregarded because "they did not work", due to unknown technical factors. In such a situation mistakes are readily made and much room is given to the preferences of the individual investigator, however critically the data may be scrutinized. The situation was not unlike that described by Ludwik Fleck for the Wassermann reaction (see chapter 4). Ludwik Fleck: "Every experimental scientist knows just how little a single experiment can prove or convince. To establish proof, an entire system of experiments and controls is needed, set up according to an assumption or style and performed by an expert" and, "the less interconnected a system of knowledge, the more magical it appears and the less stable and more miracle-prone is its reality, always in accordance with the thought style of the (thought) collective."

The undisputed leader in the development of the thought style on T cell suppression was Gershon, who was an enthusiast in T cell subsetting but did not like the idiotypic network theory. The obsession with subsetting resulted in an ever increasing number of T cell subsets that participated in suppression. The term "circuit" was coined to describe a chain of inductive events between T cell subsets, each with its own unique surface phenotype, resulting in feedback suppression[15,16]. The physiological purpose of a circuit was the fine tuning of suppressive activity exactly as required for an immune response to deal optimally with an aggressive antigen, and at the same time to avoid overshooting immunity, possibly resulting in allergies and autoimmunities. Circuits were organized in multiple layers, which were activated in succession depending on the dose of antigen and the strength of the immune reaction. A circuit began with an inducer cell which was activated by the antigen and created a signal that in turn activated a transducer cell. The activated transducer cell then passed the signal on to the suppressor effector cell which did the actual suppression. The final target was the helper T cell whose activity was regulated by the circuit, thus fine tuning all helper T cell dependent immune responses including antibody formation, cytotoxicity, macrophage activation, etc. This being not enough, the entire circuit was counteracted by a contrasuppressor circuit, also consisting of inducer, transducer, and effector cell subsets, each with its own surface phenotype, and regulating the activity of the suppressor circuit. With increasing strength of the immune reaction a next layer of suppressor and contrasuppressor circuits was activated, and so on. Gershon: "Are there more levels? Perhaps, but studies on relatively simple antigen challenges may fail to reveal these additional complexities. As we increase our understanding of immune responses to tumors and parasitic infections, however, we might well find that such levels exist"[16].

Helper and killer T cells were known to recognize their antigens in an MHC-restricted fashion, with specialized restriction elements for each of the two T cell types, corresponding to Class I antigens for killer T cells and Class II antigens for helper T cells. If suppressor T cells were such an important class of T cells, should they not also have their own restriction element in the MHC? This restriction element was soon identified, became generally accepted as a scientific fact for a number of years, yet was later shown to be non-existent. This prominent example of a collective scientific error will be discussed here in some detail: Knowledge about the mouse MHC, also called the H-2 complex, had advanced to a state where a number of subregions could be distinguished. Two subregions at either end of the gene complex contained the K and D genes that encoded Class I antigens. The central section, also called the I region, contained several genes that encoded the Class II antigens, with two genes required for each of the two heterodimeric class II molecules, I-A and I-E. The order of genes, KAED, had been established mainly by crossing congenic mice that differed only in their allelic forms of the H-2, so-called H-2 haplotypes, and by identifying offspring that had undergone crossovers between the different H-2 subregions. Inbred strains were generated that carried the recombinant haplotypes in homozygous form, and large numbers of these so-called recombinant inbred strains were kept in immunological laboratories for genetic studies on the H-2 complex, which proved of utmost value in the early research on the mouse MHC. Two such recombinant strains, termed 3R and 5R, initially appeared quite similar as they shared the allele combination derived from the two parental haplotypes, suggesting the recombinations to have taken place between A and E: K^bA^b // E^kD^d. However, it seemed theoretically unlikely that the two independent crossover events had occurred in exactly the same position. Speculating that the putative region between the two crossovers might contain a gene that encodes an antigen, two laboratories decided to produce antisera by immunizing 3R mice with 5R cells and vice versa[17,18]. Both obtained antisera that reacted with cells of the strain used for immunization, so that part of the H-2 complex was obviously different between 3R and 5R mice. They concluded that the two crossovers must have occurred at different positions, defining a new subregion between A and E, termed J, which differed between the two recombinant haplotypes: $K^bA^bJ^b/E^kD^d$ for strain 3R and $K^bA^b/J^kE^kD^d$ for strain 5R. The J region apparently encoded a novel H-2 antigen, termed I-J. One of the two laboratories subsequently demonstrated that the I-J antigen was specifically expressed on suppressor T cells, the other found that H-2 homology was required for suppressor cells to exert their function, and the H-2 subregion involved was identified as the J region. This established the I-J molecule both as a surface marker and a restriction element of suppressor T cells[16,19].

The discovery of the I-J molecule as restriction element established the suppressor T cells equal in rank to the traditional helper and killer T cells, i.e. they, too, had a special class of MHC molecules directing their function. In subsequent research I-J then became associated with nearly all suppressor T cell subsets and functions, for example in Gershon's level 1 suppressor circuit both the inducer and the transducer T cells expressed I-J as a surface marker whereas the effector cell did not express I-J but its suppressive function on helper T cells was restricted by I-J. Most notable, however, was the association of I-J with the soluble recognition molecules most suppressor T cell subsets apparently secreted[16,19]. Since the early 1970s the notion had developed that T cells, similar to B cells, exerted some of their functions by secreting antigen-specific molecules, the so-called "factors". This was first shown for helper T cells, which secreted factors which attached specifically to the immunizing antigen and appeared to be encoded by H-2 I region genes, a notion apparently influenced by the Ir-gene paradigm which had postulated antigen-specific T cell receptors encoded by MHC-linked Ir genes[20,21]. Helper T cell factors, which were later revealed as fraud, never attracted many laboratories as a subject of interest, whereas the subsequently discovered suppressor T cell factors became very fashionable and a number of leading immunological laboratories engaged in their study[22]. Nevertheless, helper factors being MHC-encoded molecules with specific antigen binding sites, set the stage for subsequent models on suppressor factors[16,19,23]. Consequently, suppressor T cell factors were found to express I-J determinants as well as specific binding sites for antigen, in different combinations and molecular versions, depending on the preferences of the investigator. For example, in Gershon's view suppressor factors consisted of two polypeptide chains, one bearing the I-J determinant and the other binding the antigen. The antigen-binding chain had an additional binding site for I-J, facilitating the formation of an active dimeric suppressor factor by an I-J restricted interaction between the I-J bearing chain and the antigen-binding chain. Several alternative models were put forward by others, all of them more dependent on phantasy than on experimental evidence.

The integration of suppressor T cell circuits into the idiotypic network was a secondary development and involved several lines of experiments. As for T cell factors, the idiotypic studies on helper T cells, while they did not raise much interest beyond a small circle of network enthusiasts, set the stage for idiotypic studies on suppressor T cells, which attracted a number of leading investigators as their main subject of interest. Similar to helper T cells[24], suppressor T cells were shown to be inducible by injecting anti-idiotypic antibodies into mice[25–27]. A network involvement of suppressor T cells, however, required a demonstration of their ability to recognize the idiotypes of helper T cells and/or B cells. Antigen-independent cooperation between idiotype-specific helper T cells and idiotype-bearing B cells had been reported for antibody responses, suggesting recognition of immunoglobulin idiotypes by helper T cells[28–30]. A pivotal influence came from a report in the early 1970s that specific suppression of the production of immunoglobulin allotypes could be adoptively transferred by T cells recovered from mice that had been treated with anti-allotype antisera[4]. As speculated by Jerne[6], if suppressor T cells recognized immunoglobulin allotypes, why not also immunoglobulin idiotypes? Following Jerne's lead, suppressor T cells were recovered from mice pretreated with anti-idiotypic antibody under certain conditions, and shown to inhibit the production of antibodies bearing that idiotype upon subsequent immunization with antigen[27]. What was still lacking now was the demonstration that idiotype-specific suppressor T cells were able to interact with other T cells by recognizing their idiotypes.

The recognition of idiotypes among T cell subsets involved in suppression was put forward with great emphasis in a series of prominently published papers, initiated in a cooperation between the laboratories of Baruj Benacerraf and Alfred Nisonoff, based on the crossreactive idiotype of antibodies to hapten ABA (azobenzoarsonate) in inbred mice[31–33]. The experiments were then carried on by Benacerraf in cooperation with Martin Dorf and Mark Greene, resulting in the characterization of a series of suppressor factors secreted by a series of suppressor T cell subsets, termed Ts1, Ts2, Ts3, and forming a circuit similar to the inducer-transducer-effector circuits described by Gershon, in this case regulating the antibody responses as well as delayed type hypersensitivity to haptens. Ts1 cells secreted a factor, TsF1, which was antigen-binding and shared the crossreactive idiotype of anti-ABA antibodies. TsF1 induced Ts2 cells to produce a factor, TsF2, which had anti-idiotypic specificity and activated Ts3 cells, which again expressed the idiotype and acted as the effector T cell in the suppressive circuit. In this model of suppression, T cells interacted with each other by idiotype-antiidiotype recognition, in full agreement with the network theory. Needless to say, some factors expressed I-J determinants, others acted in I-J-restricted fashion. Importantly, all of these interactions depended on the sharing of the immunoglobulin heavy chain allotype by all interacting partners, as would be predicted if T cell receptors, including that of suppressor T cells, utilized VH gene products for recognition of idiotypes or antigens[34].

Similar to the ABA suppressor circuit, some suppressor T cell factors in Gershon's circuits also showed VH restriction[35,36], a fact that anybody else at the time would have explained by the VH nature of T cell idiotypes. Gershon however, who worked mostly with sheep erythrocytes as antigen for which no crossreactive idiotype was known, did not like the idiotypic network. In order to find alternative explanations for the VH restriction of suppressor factors, Gershon and coworkers went to the most bizarre experiments and arguments. For example, they started working with an antiserum prepared against a tumor antigen that appeared to be encoded by a gene in loose linkage to the heavy chain genes on mouse chromosome 12. When they found that the antiserum blocked the activity of a VH-restricted T suppressor factor, Gershon suggested that the target structure of the suppressor factor was this tumor antigen rather than VH, and that VH restriction was a misinterpretation owing to the linkage between the genes encoding both structures[16,37]. Now, as we know that all of these notions had been products of imagination, it is clear that the investigator's decisions for or against the idiotypic network in T cell suppression were a matter of personal preference, not of evidence.

The year 1982 was a turning point in suppressor cell research, kicked off by reports from Leroy Hood's laboratory in which it was shown by molecular genetic methods that a gene that encodes I-J does not exist in the I region of the mouse H-2 complex, and that the H-2 complexes of strains 3R and 5R were identical[38,39]. The news hit the suppressor community like a sledgehammer, particularly as only two years earlier the first I-J bearing suppressor T cell hybridomas and the first monoclonal anti-I-J antibodies had been made, drastically boosting the credibility of I-J[40,41]. Investigators engaged in I-J related research reacted differently to this shocking result. Jan Klein wrote: "We should not go on using J as a marker for this and that, talking further about J restriction, and characterizing further factors with all kinds of reagents. We should try, instead, to see whether there is anything to J at the molecular level. Until we know that, we should perhaps be quiet about J"[42]. Others tried hard to rescue the credibility of their past work by putting forward various theories how an I-J molecule could exist without a gene in the MHC but still apparently controlled by the MHC[42]. One proposed mechanism was that I-J was a T cell recognition molecule under the selective influence of the MHC but encoded by a gene elsewhere. Others suggested that two genes are required for I-J expression, a regulatory one in the MHC and the structural one elsewhere. No matter how good these arguments were and how well they were supported by evidence, nothing helped, the suppressor field had suffered a first severe blow.

The disappearance of I-J, disastrous as it was, did not immediately kill suppressor T cell research, which survived for about another five years but with great difficulties. Moreover, the destruction of I-J was not the only reason for the growing discredit of suppressor cell research that followed. Already before there had been an increasing feeling of uneasiness among immunologists about several aspects in the field, including the excessive T cell subsetting, the complexity of the arrays of factors involved, the differences between the results coming from different laboratories, etc. In addition, those who tried presumably experienced difficulties to reproduce certain experiments in their own laboratories. As a result, manuscripts were more frequently rejected by the peer-reviewed journals. Gershon, who as the leading character in the field suffered perhaps more than others from the difficulty to publish, founded a new immunological journal, entitled *Journal of Molecular and Cellular Immunology*, of which he became the first editor-in-chief. The editorial process included peer review but, even in case of harsh criticisms, the author could insist on the publication of the manuscript. In this case, the critiques were published together with the paper, with reviewers free to choose having their names disclosed or not. Authors could choose to write a rebuttal which was also included. Although it seemed an interesting addition to the existing array of journals, *Journal of Molecular and Cellular Immunology* never made it into the top class of journals. Apparently the argumentative style, which presented differences of opinion unresolved to the readership, stood against the inherent desire of scientists to produce, and get exposed to, definitive results. After Gershon's death in 1983 Charles "Charly" Janeway Jr took over as editor-in-chief, but the journal did not survive for more than a decade.

In the late 1980s a growing consensus developed among immunologists to stop working on, and talking about, suppressor T cells. As a result, suppressor T cells virtually became banned by granting agencies, journals, conference schedules, etc. An important role had an editorial entitled "Do suppressor T cells exist?" written by Göran Möller and published in the *Scandinavian Journal of Immunology* in 1988[43]. In addition to the disappearance of I-J and the fuzziness of the various cellular subsets and their factors, it had in the meantime also been shown that suppressor T cells and hybridomas, although they displayed antigen specificity, did not posses or express rearranged T cell receptor genes[44]. Möller's timely editorial summarized these and other shortcomings of suppressor T cell research sharply and mercilessly. In subsequent issues several scientists, involved or not in suppressor cell research, commented on Möller's points, including myself. Having been involved in suppressor T cell research and convinced of the existence of T cell suppression, I described my position as follows: "...neither the absence of a marker, nor the lack of appropriately rearranged (T cell receptor) $\alpha\beta$ genes, nor the failing basis for the existence of I-J provide proper arguments to deal with the existence of suppressor cells. They do provide arguments, however, for the validity of previous concepts of suppression. It is in this context that I agree with Dr. Möller: Many previous concepts, particularly those composed to be played by the famous Gershonian philharmonic orchestra (a metaphor frequently used by Gershon himself), have failed to provide a useful background for continuing advance in the elucidation of the cellular and molecular basis of suppression". My commentary[45] had been entitled: "Suppression needs a new hypothesis."

It took about a decade before suppressor T cells became acceptable again in immunology, now under the term "regulatory T cells", or "Treg"[46,47]. The "new hypothesis" on suppression, however, goes back to a line of experiment started as early as 1969, predating Gershon's first report, and having developed largely unnoticed in the shadow of mainstream suppressor cell research. Japanese researchers had reported that newborn mice whose thymus was removed shortly after birth developed an atrophy of the ovaries[48]. First interpreted as lack of a thymic hormone required for proper ovarian development, subsequent investigations into the phenomenon revealed it to be an autoimmunity, not restricted to ovaries but affecting multiple mostly endocrine organs. Neonatal thymectomy thus caused generalized autoimmunity, suggesting that the neonatal thymus produced a component inhibiting the development of autoimmune reactions. This component was then identified as a T cell, Treg, released by the thymus after birth into the peripheral lymphoid tissues and expressing the markers CD4 and CD25[49]. Helper and cytotoxic T cells are released by the thymus already before birth, containing some autoaggressive cells that escape thymic selection. In the normal animal, these autoaggressive cells are suppressed by the Treg, whereas in the neonatally thymectomized animal autoimmunity develops. Also when isolated from the spleen or lymph nodes of adult mice and given to neonatally thymectomized mice, Treg prevent the development of autoimmunities. Treg can also be induced by special experimental immunization procedures and their suppressive function can be demonstrated in immune responses to experimental antigens. Heritable diseases are known in mouse and man in which Treg are missing, resulting in generalized autoimmunity[50]. The defective gene in the mouse has been identified as encoding a transcription factor that is specifically expressed in Treg, FoxP3, which so far seems to serve as a specific molecular marker. Treg represent a single subset which expresses CD4 rather than CD8 (formerly Ly2/3) and so are not rediscovered versions of the former suppressor T cells. Treg express rearranged T cell receptor genes, are restricted to class II MHC antigens, and no soluble antigen-specific factors are involved in their function.

Chapter 12
Network mannerism

For the choice of their artistic means, the artists had at their disposal the entire spectrum of degrees of reality from naturalism to idealism, they had the possibility to create out of phantasy and to place realistic elements side by side with stylistic ones, to combine the disparate and to mix the classical with the abstruse.

Erwin Lachnit (1987) History of the Notion of Mannerism.
In: *Zauber der Medusa, Europäische Manierismen.*
Löcker Verlag, Wien

Artists of the Italian Renaissance developed their characteristic styles from the observation of nature and the formulation of a pictorial science. In 1520, when Raphael died, all the representational problems had been solved, painting had been established as a craft to be learned. The artists who followed, however, instead of taking nature as their teacher, took art itself. This epoch is referred to as "mannerism", because it had turned style into manner. Today it is realized that most styles of art are followed by a period of mannerism, in this context often referred to as "post-modernism". Umberto Eco, esteemed author of works of science as well as of fiction, wrote: "I have come to believe that "postmodern" is not a movement confined to a particular time, but a state of mind or, more precisely, an approach to intentional art. One could even say that every epoch has its own post-modernism, just as it was said that every epoch has its own mannerism"*.

I have come to suspect that not only styles of art have their mannerism, scientific thought styles may have theirs as well. Certainly, the network the-

---

* Quoted from Eco U (1984) Postscript to *The Name of the Rose*. Hanser, Munich/Vienna, translated by the author.

ory experienced a post-modern, manneristic period in which the network theory itself became the object of interest, rather than the immune system it was meant to describe. In this context, humanists joined the field and immunologists, too, began to discuss the network theory using jargon and thought styles adopted from philosophy, sociology, linguistics, etc. As a consequence, the network paradigm became to an extent detached from its roots in basic biology, however weak they may have been. Philosophical notions such as semiosis or autopoiesis began to conquer the field, much at the expense of biological meaning and scientific reason. Jerne himself joined this development. In an article of 1985 entitled "The generative grammar of the immune system"[1] he drew a parallel between immunity and language, and between immunology and linguistics, quoting extensively from the linguist, Noam Chomsky[2]. Jerne proposed similar principles of information engraved in the amino acid sequence of the combining site of an antibody and in the sequence of words in a spoken sentence. Jerne must have noted himself that by such comparisons the issues at stake became more cloudy than clear: "No matter what you try to investigate in biology turns out to become increasingly complex". In another article, entitled "Idiotypic networks and other preconceived ideas"[3], he referred to the cybernetic artist Nicolas Schöffer's book *La Théorie de Miroirs*[4], and recommended that "...those who always seek exterior pressures (e.g. microbes) to account for the evolution of the V genes, would do well to turn their vision towards the interior of themselves, and there discover the mystery, perhaps never completely revealable, of the immune system". The immune system, rather than being a – however difficult to tackle – biological system of cells and molecules, was so declared a mystery.

One of the initial motions towards the development of network mannerism was the coining of the term "complete" by Antonio Coutinho, to describe the size of the repertoire of recognition structures in the immune system. Before, immunologists had strained their minds, multiplying not precisely known numbers of multiple immunoglobulin gene segments with uncertain rates estimated for somatic mutations and nucleotide insertions, to arrive at very high but frustratingly imprecise estimates for antibody or T cell receptor repertoires. Now, Coutinho made the radical decision to call the repertoire complete[5,6]. The original paper of 1980[5], a review article in which the immune system was also attributed as "promethean", was exceptionally prolific. Although the paper was directly cited only about 50 times, no subsequent conceptual treatise of the immune system could get by without referring to the completeness of the repertoire. Jerne himself was no exception.

For those who had always been critical of the network thought style the notion of completeness was even more infuriating. For example, Melvin Cohn, together with Peter Bretscher author of the associative recognition theory[7], the only existing alternative theory of immunity, felt obliged to publish a comment on completeness, together with Rod Langman. In their comment, entitled "The 'complete' idiotypic network is an absurd immune system"[8], they argued that the term complete could only mean infinite, and since a mouse only had about 10^8 lymphocytes, each with only one type of receptor, its receptor repertoire cannot be infinite but must be finite. They did not forget to point out that the notion of crossreactivity does not help in rescuing infinity, it only transfers the problem to a single combining site with an infinite number of specificities, i.e. specificity taken ad absurdum. Impeccably logical as Cohn's and Langman's rejection of completeness was, it could not hamper the popularity of the term. What Cohn and Langman did not understand was that the term completeness did not convey, and was not meant to convey, any exact meaning in the scientific sense. Rather it was meant to confer a nebulous impression of encyclopedic almightiness. Everything, including all exogenous epitopes and all endogenous idiotopes, was anticipated in the immune system, just as human needs and wishes were anticipated by the clairvoyant Prometheus, creator and benefactor of mankind in antique mythology.

The term completeness also implied a radical dismissal of the classical concept of self-nonself discrimination as a fundamental property of the immune system. Since Paul Ehrlich's notion of "horror autotoxicus"[9], receptors for self-epitopes had been forbidden in the immune system and generations of immunologists had tried to unravel how this was achieved, by clonal deletion, somatic mutation, anergy, suppression, no possibility was left out. A complete repertoire, in contrast, did not exclude receptors for self[6]. Receptors were selected in ontogeny to recognize self-idiotopes, so why not other self-epitopes as well. Only because idiotopes far outnumbered any other structure in the body did it seem that self-recognition was primarily idiotypic. Immune reactions to exogenous epitopes were merely unavoidable accidents resulting from the completeness of the repertoire. The fact that the immune system was usually tolerant of self was no longer due to deleted or anergized clones with anti-self receptors. In contrast, tolerance of self was a property of the network, which in some unknown way managed to keep clones with anti-self receptors at bay. There were some modelling attempts to unravel the ways of tolerance in an idiotypic network, mathematicians had calculated that a network can have stable states in which nothing happens, so this was what might be the case in tolerance. Nevertheless, the network was too complex to completely understand all of its intricacies. In the end, it remained a mystery, no matter how hard you tried.

Doing away with self-nonself discrimination had even more drastic consequences for immunology: It meant doing away with what had previously been taken as the sole evolutionary cause for the existence of the immune system, namely the recognition of and protection against foreign pathogenic microbes and their toxic products. Nelson Vaz and colleagues, in trying to delineate an alternative evolutionary cause for the immune system[10], argued using the mechanism of "exaptation", a term introduced by S.J. Gould when discussing the spandrels in architecture as an evolutionary paradigma[11]. A spandrel is a triangular structure that stabilizes arches in medieval churches and palaces. Spandrels were later mostly used as scaffolds for displaying rich decorations and ornaments, thus acquiring a function essentially unrelated to their original purpose. For such an opportunistic role change Gould coined the term "exaptation", in contrast to the term adaptation which denotes a change to a novel function that serves a somewhat altered but otherwise conserved purpose. Vaz and colleagues argued that molecular precursors of many of the constituents of the immune system of higher animals had served unrelated biological functions in lower animals. For example, the homologues of certain highly conserved Toll-like receptors, signaling molecules, or transcription factors, which are now essential elements of the immune system, serve non-immune functions in insects or worms. Since some of these molecules have been acquired from microbes or plants by horizontal gene transfer, their change in function was not by adaptation to gradually evolving needs but by fortuitous exaptation. Accordingly, Vaz and colleagues proposed that the protective functions of the immune system were a fortuitous accident: "Defense against infectious disease is a most powerful aim to be fulfilled, and it is fair to assume that contemporary organisms ... are equipped with an array of receptors which ... is vast enough to cope with the invasion of microorganisms. Defense, however, seems to be more of a side issue to the more important issue of molecular identity, of which the immune system is the essential regulator"[10].

The immune system thus became the "essential regulator of molecular identity". The regulator, also referred to as the "agent", was the idiotypic network, from which properties "emerged" that were far more complex and competent than was expected from the sum of its clones, would they function independently. The network functioned in the way of the "cognitive domain" of the central nervous system, i.e. there was an interdependence between the characteristics of its own elements and the domain of its possible interactions. Translated for the immune system this meant that its interactions were determined essentially by its own elements: "...Whatever may be recognized as an antigen may be so recognized because it has a degree of resemblance to an already existing set of determinants in the network – an internal image...". An antigen was an antigen only when it resembled a paratope or idiotope in the immune system. The outside world of dangerous antigens had thus been degraded from representing the evolutionary necessity for the immune system in higher organisms to a set of accidental similarities with some endogenous structures designed to form a network. Essential criteria of the network were its "Eigenbehavior", its "closure", its "recursion", and its "self-reference", i.e. it formed a closed, self-sufficient system that created its own boundaries, observed its own condition and decided in its own right to make adjustments if this deemed appropriate. Protection against infectious agents was an occasional benefit of no more than marginal interest[10,12,13].

The concept of molecular identity as the evolutionary cause of the immune system was carried further by Francesco Varela, biologist, epistemologist, scholar of cognitive science, and an international intellectual celebrity[13]. Together with Humberto Maturana he stands for the concept of "autopoiesis"[14], a novel philosophical approach to living systems initially developed for the cell: "Autopoiesis attempts to define the uniqueness of the emergence that produces life in its fundamental cellular form. It's specific to the cellular level. There's a circular or network process that engenders a paradox: a self-organizing network of biochemical reactions produces molecules, which do something specific and unique: they create a boundary, a membrane, which constrains the network that has produced the constituents of the membrane. This is a logical bootstrap, a loop ... A self-distinguishing entity exists when the bootstrap is completed.... It doesn't require an external agent to notice it, or to say, "I'm here." It is, by itself, a self-distinction. It bootstraps itself out of a soup of chemistry and physics"[13].

Autopoiesis belongs to the second order cybernetic philosophies which, in contrast to first order cybernetics which focussed on negative feedback regulation and stability in mechanical systems, stressed the importance of positive feedback, growth, instability, and the unpredictable "emergence" of novel entities as essential properties of living systems. Life is thus "An emergent property, which is produced by an underlying network, is a coherent condition that allows the system in which it exists to interface at that level – that is, with other selves or identities of the same kind. You can never say, 'This property is here; it's in this component'. In the case of autopoiesis, you can't say that life – the condition of being self-produced – is in this molecule, or in the DNA, or in the cellular membrane, or in the protein. Life is in the configuration and in the dynamical pattern, which is what embodies it as an emergent property."[13]

Varela applied the principles of autopoiesis also to multicellular systems including the central nervous and immune systems. For the latter, Varela interacted with Vaz and Coutinho, resulting in several individual and collaborative articles in which the three set about to revolutionize immunology[12,15]. Varela: "...classic immunology understands immunology in military terms – as a defense system against invaders. I've been developing a different view of immunology – namely, that the immune system has its own closure, its own network quality. The emergent identity of this system is the identity of your body, which is not a defensive identity. This is a positive statement, not a negative one, and it changes everything in immunology.... We have to go beyond an information-processing model, in which incoming information is acted upon by the system. The immune system is not spatially fixed, it's best understood as an emergent network".

As an example for his view of the role of the immune system in infections, Varela communicated these thoughts about AIDS[13]: "AIDS is a dramatic case of the deregulation of this coherent emergent property, much like ecological dysfunctioning. People think AIDS is an infection. This is, of course, true, but not true in the sense that once the system is infected with AIDS it triggers a condition of self-destruction of the immune system. HIV triggers a deregulation, which then amplifies itself and becomes its own nightmare." Moreover: "This is typical of an autoimmune condition: the system eats itself up. Consequently, it's beginning to dawn on people that looking for AIDS vaccines is a complete waste of time. From my point of view, the right approach is first to understand the nature of this global regulation. One hint of how to do this is to look for ways to reconnect the system.... One approach we study is to provide new, normal antibodies that help to re-create the network." In other words, with an intact network not even AIDS could do you any harm.

What may have been the ultimate highlight of network mannerism was an attempt to apply the linguistic theory of semiotics to the immune system. Semiotics as the theory of the function of signs in communication arose as a subspecialty of linguistics, trying to analyze how information is passed on between humans talking one to another[16,17]. In semiotics words are signs which "mean what they mean by virtue of the fact that they are not the thing that they mean". In very basic terms, the semiotic process of communication involves three subjects, the sign, its object, and its interpretant. For example the word (sign) "automobile", the object automobile (a shiny new Mercedes), and the interpretant: Jee, he just bought a Mercedes, he must be rich! Important is that, dependent on the context, for the same sign (word) there are many alternative objects and even more interpretants. To stay with the example, the object may be a car damaged by a traffic accident and the interpretant may be: Damn, this repair will be expensive! And so on.

Applications of the concepts of semiotics to communicative systems other than language are manifold, including the communication among the cells and molecules of complex living organisms. Umberto Eco, professor of semiotics at the University of Bologna and widely known as author of the novel *The Name of the Rose*, wrote: "...for a long time semiotics has been considered as an imperialistic discipline aiming at capturing every aspect of the world"[16]. A protagonist of bio-semiotics was Thure von Uexküll, who wrote[18]: "If one reflects upon the fact that the human body consists of 25 trillion cells, which is more than 2000 times the number of people living on earth, and that these cells have direct or indirect contact with each other through sign processes, one gets an impression of the amount. Only a fraction are known to us. Yet this fraction is hardly comprehensible.... The *messages* that are transmitted include information about the meaning of processes in one system of the body ... for other systems as well as for the integrative regulation systems (especially the brain) and the control systems (such as the immune system)."

The idea to bring immunologists and semioticians together in a scientific conference was born in a conversation among Eli Sercarz, Franco Celada,

Avrion Mitchison, and Tomio Tada, when they were discussing the subject of T cell-B cell cooperation on the way to a meeting in California in 1984. When Tada threw the word "immunosemiotics" into the discussion they agreed to collaborate towards organizing a meeting on the subject. Celada happened to know Umberto Eco, and was asked to contact Eco with the aim to determine if he and his collaborators were willing to attend an interdisciplinary conference together with immunologists. They were, and the meeting took place two years later in a mountain resort called Il Chicco near Lucca in Tuscany. The proceedings were published in 1988[19], a most revealing document of scientific post-modernism.

The immunologists invited to the conference had been selected because of their previous conceptual contributions to the network paradigm and, in order to acquire some basic knowledge on semiotics, had been asked to read two of Eco's scientific works[16]. As a result, many of the immunologists did their best to find some link between the immune system and semiotics. Here are some quotes: "...it follows that the same receptor recognizes and responds to both the antigen and the "sign" of the antigen, the internal image. From this it can be seen that the network is its own sign system which sees in itself the sign of the antigen" (Ed Golub). "The message in immunosemiotics is rather like a text since it represents a set of signals which are connected. As in language, one sign can be replaced by another and therefore can induce a set of different responses. The major characteristic of the interpretant (i.e. the lymphocyte) is that it is able to specifically select a definitive information input because of the specificity of its receptor..." (Constantin Bona). "All in all, it would thus appear that the antigen-specific cooperation between T cells and B cells is based on one principle; that when confronted with the same sentence (antigen "A"), both antigen-presenting macrophage and anti-antigen "A" B cells chose the same word (a particular amphipathic α-helical segment) out of that sentence..."(Susumo Ohno).

When listening to the semiotician's lectures it was evident that most of them were rather unwilling to grant a semiotic quality to the immune system. Eco was rather outspoken: "In semiosis the criteria for recognition change according to different contexts. Can immunologists say that the same happens with their cellular pets?", and "Immunologists frequently use the word recognition. I walk, I put my foot upon a hole in the ground and I stumble. Would I say that I 'recognize' the hole? I think that most of the steric phenomena considered by genetics belong to this kind of mere stimulus-response process." Finally he reacted with some good sense of humor: "At the beginning of our meeting ... I felt the duty of stating many reasons that ... discouraged a direct application of semiotics to immunology. After the meeting I have changed my mind. I feel still unable to say whether semiotics can help immunology, but I discovered that immunology can help semiotics.... By a happy case of serendipity, our meeting did something for the advancement of learning." He contributed an introductory article as well as a series of doodles to the proceedings of the meeting, but neither disclosed the ways in which immunology could help semiotics.

Other participants presented their long-standing ideas in immunology or semiotics much without reference to the other field. For example, Coutinho, with F. Jaquemart, elaborated on the untenability of self-nonself discrimination: "If we refer to common usage, it seems that 'self' is conceived as a set of de jure listable entities.... Non-self, in contrast, is the negation of self.... However, are this self and non-self in the same terms? We will not normally say that a cloud or a lake is non-self for the mouse, but not because it is wrong. Focusing on an entity (i.e. the mouse), when speaking about self implies at once, TO SOME EXTENT, a sort of positive constitution of the non-self domain. Nothing FORMALLY prevents us from considering a screw as mouse non-self, but, in immunology, we do not expect to get anti-screw antibodies..... Focussing on the self, VERY ROUGHLY, determines the domain of relevance in its negation. This implies that we cannot use here the very classical postulate: non non-self = self.... Non-self being negatively constituted, it could be a set of de jure listable entities only if one could postulate that non non-self = self and if the self was represented by a set of DE FACTO listable entities (and not only de jure)." (capitals by the authors).

Many participants of the conference were rather disappointed with its achievements or lack thereof. Klaus Rajewsky brought it to the point when he said: "I was hoping that one could understand immunology on a higher level". The proceedings of the conference were published when the network paradigm was already on its way down, and the attention they received was close to nil. Eli Sercarz: "I like the book, particularly Umberto Eco's doodles".

Chapter 13

Post-network immunology: Idiotypic network continues at the bedside

I don't think that there is a problem here. Sometimes techniques and devices developed in the laboratory move into our larger environment and indeed help us in some already-chosen mission.
Ian Hacking (1992) The Self-Vindication of Laboratory Sciences. In: A Pickering (ed) *Science as Practice and Culture.* The University of Chicago Press

As reflected by the quotation rates given in Figure 9.4 (chapter 9), from 1990 onwards interest in the idiotypic network theory rapidly vanished, particularly from basic immunological research. Nevertheless, until today it enjoys a steady rate of about 20–40 quotations/year, reflecting continuing interest by clinicians and medical researchers working in companies.

In basic immunology the network paradigm became replaced by several others, in a fashion more drastic and complete than the Kuhnian paradigm shifts that are part of the normal scientific progress. In a paradigm shift certain elements of preexisting knowledge and tradition are discarded while others are maintained, leading to a developmental leap in a field that nevertheless maintains much of its previous foundations. In contrast, the network paradigm became, for a time, almost totally discredited as misleading, non-productive, even non-scientific and imaginary in all of its parts. Instrumental in this process were two developments that were already discussed, the unmasking of the immunoglobulin nature of the T cell receptor as a fallacy (see chapter 10), and the disgrace of suppressor T cells triggered off by the loss of the molecular basis for I-J (see chapter 11). The crises caused by both of these developments, which took place around 1984 and 1988, respectively, together caused a turmoil that severely compromised the network paradigm even though neither had constituted a central network element. The turmoil was heightened by a growing irritation in the scientific community with the various attempts to mystify the idiotypic network as a quasi philosophical paradigm (see chapter 12). And not only the network paradigm became compromised. It was the entire traditional approach to immunology, in which theoretical considerations often dominated over experimental evidence, that became discredited. Ever since, the network theory has been essentially banned from the conceptional discourse in immunology.

Indeed, the disappointment with the network theory brought with it a general attitude of mistrust and reluctance towards comprehensive theories as such. As a result, no unifying theory of adaptive immunity has replaced the network theory until today. Rather, post-network immunology is conducted under a handful of coexisting paradigms, each covering a particular sector of adaptive immunity, but falling short of forming a cohesive picture. Most of these small paradigms developed as a result of a bunch of novel technologies that had revolutionized the field from the mid 1970s on, including monoclonal antibodies and flow cytometry, T cell cloning and hybridomas, DNA/RNA sequencing and cloning, analytical peptide and protein chemistry together with a quantum leap in the sophistication of x-ray crystallography, and, last not least, transgenic and knockout mice. Such a host of novel techniques invited a new, technology-oriented approach in immunology, followed by a flood of molecular information which, to no-one's surprise, consisted mostly of unconnected results that only slowly integrate into coherent concepts. At present, immunology appears to be in a state in which some of these concepts gave rise to paradigms that govern the sectors that had lent themselves particularly well to the application of novel technologies. Other sectors, not so readily approached by the new methodologies, remained as hazy as before. Another attempt at a comprehensive theory of adaptive immunity, replacing the network theory, has so far not been made. Only some of the conceptually pivotal post-network paradigms are roughly sketched below.

The dominant role of gene technology in immunological research began as early as 1976 when Hozumi and Tonegawa first showed that by studying DNA one can obtain unprecedented insights into the generation of antibody genes by somatic recombination[1]. This and a second report of the same group in 1978, showing that V genes come in pieces[2], gave rise to a novel paradigm that governed more than two decades of research on the somatic generation of antibody diversity, a sector of immunology which today is almost completely understood. The molecular work on the T cell receptor opened the view on the second system of diverse receptor molecules in the immunology, similar in design to immunoglobulins but encoded in an unrelated gene complex[3]. In 1984/5 the first transgenic mice were produced expressing individual antibody polypeptide chains or complete antibody molecules encoded by cloned antibody genes artificially inserted into the mouse genome[4]. In 1990, the first report appeared on the artificial deletion of a gene of interest in immunology, the β2-microglobulin gene, from the mouse genome by homologous recombination[5]. The "knockout" mouse paradigm was born, an unprecedented success story that eventually provided biologists with the opportunity to assess the vital functions of virtually any mammalian gene product. Immunological genes were uniquely amenable to this approach, as most of them were not essential for viability, their deletion only causing a "phenotype", namely immunodeficiencies. Knockout phenotypes were particularly instrumental in the present detailed molecular knowledge on lymphocyte development, receptor repertoire selection, signal transduction, and cell activation. In combination with the human genome, knockout mice have helped to identify the gene defects accounting for a multitude of genetically determined human immune disorders.

Transgenic mice expressing antibody or T cell receptor genes in combination with genes encoding various natural or artificial self-antigens uncovered clonal deletion as a major mechanism of self-tolerance[6,7]. As shown in 1988/89, for T cells more strictly than for B cells, clones that recognize self-antigens are deleted by negative selection[7]. In view of the importance of self-tolerance for survival, the immune system engenders several backup mechanisms, including various types of clonal anergy as well as regulatory T cells, in case a self-reactive clone has escaped deletion. Nevertheless, clonal deletion remains the predominant route to self-tolerance, at least as far as T cells are concerned. A putative role of an idiotypic network in self-tolerance, if any, may only be marginal. The clonal deletion paradigm re-established self-nonself discrimination, which had been eliminated by the network theory, as a pivotal principle of adaptive immunity.

IL-2 was the first immunological cytokine to be discovered in 1972[8], and to be cloned in 1983[9]. In the meantime and thereafter, more than 30 different cytokines were discovered and characterized, and continue to be discovered every year. Cytokines are soluble mediators secreted by certain cells and regulating the activities of other cells bearing appropriate cytokine receptors. They thus directly competed with the idiotypic network for the domain of immune regulation, immune regulation by cytokines being independent of antigen receptors and antigen non-specific. Initially the network was more successful, as it was difficult to make sense of the large number of mediators with partially different and partially overlapping cellular origins, cellular targets, and functions. In 1986, Tim Mossman and Robert Coffman discovered that T helper cell clones could be divided into different types on the basis of their secreted cytokines[10]: One type, termed Th1, secreted interferon-γ which activates macrophages, another type, termed Th2, secreted IL-4 and IL-5 which stimulate B cells to produce antibodies. These findings correlated strikingly with the historical dichotomy of cellular *versus* humoral immunity: Intracellular pathogens, for example tubercle bacilli, stimulate a cellular immune response resulting in the killing of the bacilli in activated macrophages; extracellular pathogens, including extracellular bacteria and cytopathic viruses, stimulate production of specific antibodies that attach to these pathogens and neutralize them. In order to be effective against a particular pathogen, the immune system must chose between a cellular and humoral immune response: Antibodies would be useless against tubercle bacilli, just as activated macrophages would be useless against cytopathic viruses. With the Th1/Th2 paradigm the first clue was at hand that these choices involve the activation of different kinds of T helper cells *via* secretion of cytokines. Although later studies revealed the situation as more complex, the original discovery of Mossman and Coffman generated a novel paradigm of immune regulation, replacing the idiotypic network.

In 1987 the first crystal structures of MHC molecules were reported[11]. They consisted of a bulky stem anchoring them on the cell membrane, a flat plateau on top of the stem, and two elongated bulges arranged in parallel on the plateau. Both class I and class II molecules shared a similar structure, with some class-specific features. Between the bulges there was a cleft which was not clearly delineable in the first crystals, suggesting molecular heterogeneity of materials contained in the cleft. Already in 1985 MHC molecules had been shown to bind peptides, suggesting that the immunological function of the MHC molecules was to present peptides for recognition by T cells[12]. In 1990 the material in the clefts of the MHC molecules was identified as a mixture of short peptides of many different kinds[13,14]. For MHC class I molecules the peptides were derived from degraded cytoplasmic proteins, for MHC class II molecules the peptides came from degraded extracellular proteins. The peptides are generated by proteolytic processing of proteins, self or foreign, inside the cell. Upon biosynthesis, MHC class I molecules are loaded with peptides from proteins synthesized and degraded in the cytoplasm, which include viral proteins. MHC class II molecules are loaded with peptides of exogenous proteins, self or foreign, which are endocytosed and degraded in endosomes. MHC-peptide complexes are transported to the cell surface where they can be recognized by T cells bearing appropriate receptors. The antigen processing paradigm offered the solution to the long standing problem of how T cells recognize sequential epitopes in proteins and MHC molecules with one receptor. Moreover, it opened a novel field of research directed at defining the protein epitopes recognized by T cells, for example in infectious agents, autoimmunities, etc.

Antigen processing and thymic selection are two paradigms that have managed to close the gap between them. In the thymus, where presentation of self-peptides predominates, weak recognition of an MHC-peptide complex is required for survival of the immature T cell whereas strong recognition results in its deletion[7,15]. Mature T cells that leave the thymus thus recognize MHC molecules, but only those that present peptides not present in the thymus, i.e. derived from foreign proteins. This is associated with another choice for efficiency: Killer T cells, equipped for killing infected cells, recognize MHC class I molecules and are thus confronted with cytoplasmic peptides that may derive from intracellular bacteria or non-cytopathic viruses. Conversely, helper T cells recognize MHC class II molecules and are thus confronted with peptides derived from exogenous antigens, including extracellular bacteria and cytopathic viruses, against which antibodies are effective. These conclusions made a host of puzzle pieces fall into place, with no necessity for network regulation.

The immune system reacts not exclusively but primarily with pathogenic microorganisms, but how does it distinguish between harmless foreign antigens and dangerous microbes? In the early 1990s Polly Matzinger postulated the existence of a mechanism by which the immune system identifies a dangerous antigen[16], and Charly Janeway directed the attention to the long standing empirical fact, so far taken for granted, that experimental immunizations had much better success if one used a so-called adjuvant to which the antigen was admixed[17,18]. Most common was Freund's adjuvant, a suspension of tubercle bacilli in oil, referred to by Janeway as "the immunologist's dirty little secret". Immunologists began to suspect that this was to do with the so-called innate immune system, a term describing a number of invariant defense mechanisms already present in invertebrates but maintained in evolution so as to coexist with the adaptive immune system in vertebrates. Because of its astounding specificity and molecular diversity, many immunologists had been attracted by the adaptive immune system, while the innate immune system, which showed little diversity and rather broad specificity, received little attention. Matzinger's and Janeway's contributions fell together with the discovery of a mutant of the fruit fly *Drosophila melanogaster* that was very sensitive to fungal infections[19], all leading to a renewed interest in the innate immune system. The mutated gene, termed *Toll*, turned out to be highly conserved and was also found, in similar form, in many vertebrates including mammals and man. The molecular structure of the Toll protein was that of a transmembrane receptor, binding a ligand on the outside and connecting to the signalling machinery on the inside of the cell membrane. The ligand was identified as zymosan, an extract of yeast cell membranes, consistent with a function of Toll in the control of fungal infections. Very rapidly about a dozen additional Toll-like receptors were detected in mice and man[20], all of which binding so-called "pathogen-associated (molecular) patterns", such as bacterial lipopolysaccharides, lipoproteins and peptidoglycans, bacterial GC-rich DNA, viral double-stranded RNA, etc. Toll-like receptors were found to be expressed on several types of cells that had long been known as "antigen-presenting cells", including dendritic cells and macrophages in the skin and lymphoid organs, because they are particularly efficient in presenting MHC/peptide complexes and activating T cells. The notion developed that dangerous microbes, if identified as such by Toll-like receptors on antigen-presenting cells, activate the antigen-presenting cell to induce a particularly vigorous T cell response, leading to either efficient antibody production, macrophage activation, or killer T cells, whatever is required. The danger paradigm continues to govern a fair amount of research in immunology today.

Although the network theory has no role whatsoever in any of the paradigms that guide mainstream basic immunological research today, it is not entirely dead, as indicated by occasional reports on phenomena that seem to demand interpretations in its terms. The most recent of such observations seems to point to a role of immunoglobulins in the development of a heterogeneous T cell repertoire[21]. Using a trustworthy molecular method to quantify the diversity of T cell receptors in the mouse, it was found that mice that lack immunoglobulins due to genetic deletion of B cells show a 10-fold reduction in T cell receptor diversity. Mice that possess normal amounts of B cells and immunoglobulins, but only of a single B cell clone, still have a similarly reduced T cell diversity, suggesting that diverse immunoglobulins are required for the development of a diverse T cell repertoire. These observations seem to corroborate previous hypotheses on the role of immunoglobulin variable regions in the selection of T cells in the thymus: After a direct coding of VH genes for T cell receptor idiotypes had become untenable (see chapter 10), the selection of the T cell repertoire on endogenous immunoglobulin variable regions had been considered as an alternative explanation[22], a notion that seems to be corroborated by these new results[21]. Interestingly, however, the investigators who report on these novel observations strictly avoid referring to the network theory, as if it were contagious.

Other than at the bench, the network paradigm survived much longer in medical research and continues to be active in clinical laboratories and at the bedside. Suggestions to apply the network concept to vaccination against infections go back to the very early experiments showing that anti-idiotypic antibodies could be used to induce antigen-specific B cell and T cell responses in mice[23]. Based on the notion of "internal image", it was argued that any foreign antigen could be mimicked by anti-idiotypic antibodies and consequently could be replaced by anti-idiotypic antibodies in immunization procedures. Indeed, induction of anti-microbial immunity by anti-idiotypic immunization was subsequently shown in a fair number of mouse models, including viral, parasitic and bacterial infections. The rationale for these attempts arose from the fact that there were, and still are, a number of infectious agents for which there are no safe and effective vaccines. These include cases in which the organism cannot be grown in culture, stable attenuated strains are not available or even the attenuated organism is toxic, or the host fails to mount a protective immune response when immunized with the organism or a subunit vaccine. A prominent example of the latter case is childhood meningitis, caused by the bacteria *Hemophilus influenzae*, *Streptococcus pneumoniae*, or *Neisseria meningitidis*. Protective immunity against the disease is conferred by antibodies to the capsular polysaccharides of these agents, but infants do not produce antibodies upon vaccination with polysaccharides. Another case is hepatitis B where 10% of the population cannot produce antibodies against the recombinant surface antigen vaccine. Intensive research went into the development of vaccines on the basis of anti-idiotypic antibodies until a few years ago. An approach that was quite successful in animal experiments employed the design of peptides, so-called mimotopes, on the basis of the amino acid sequences of the hypervariable regions of anti-idiotypic antibodies. For example, such peptides, coupled to appropriate immunogenic carriers, were shown to induce protective antibodies to phosphorylcholine of *Streptococcus pneumoniae*[24,25] and to the hepatitis B surface antigen[26]. Intensive research went also into idiotypic approaches to vaccination against AIDS, mostly based on anti-idiotypic antibodies against anti-CD4 antibodies, and aiming at the induction of antibodies that block the CD4-binding site of HIV gp120[27,28]. Nevertheless, in spite of all these reasonable considerations and intriguing animal experiments, no anti-idiotypic vaccine has made it to the stage of human application against an infectious disease. While until 1991 a fair number of major review articles appeared on the subject[29], only a handful of original papers and the occasional review[24] dealt with the subject thereafter.

In contrast to vaccination against infection, idiotypic approaches seem to remain quite promising in the treatment of certain cancers, particularly human non-Hodgkin B cell lymphoma. These monoclonal tumors often express surface immunoglobulin so that idiotype seems to be a reasonable target for therapy. As first shown for mouse myelomas in 1972, injection of anti-idiotypic antibodies can protect mice from the transplanted tumors[30]. In 1982, Ron Levy and coworkers were the first to report the effects of treating patients with B cell lymphoma by injecting custom-made anti-idiotypic antibodies[31]. This approach of passive serotherapy, in addition to the cumbersome and time-consuming production of a monoclonal anti-idiotypic antibody in each case, had a number of features that undermined its success, including reduced or loss of immunoglobulin production by the tumor cells, or alteration of idiotype. Levy and coworkers then turned to developing the technical means for active immunization of patients against their lymphoma idiotypes. The lymphoma cells are first hybridized to generate hybridomas that produce the lymphoma immunoglobulin in large quantities. The purified immunoglobulin is then coupled to an immunogenic carrier protein and patients are repeatedly immunized with this complex, using a hematopoietic cytokine, GM-CSF, as adjuvant. Patients treated in this way generate idiotype-specific humoral and cellular immunity against their tumor cells, which seems to extend to idiotypic mutants as well. In a recent clinical study on patients with follicular lymphoma, a particular form of non-Hodgkin B cell lymphoma, it was shown that idiotypic vaccination significantly prolonged the duration of remissions after combined chemotherapy[32,33]. The result was hailed as "unprecedented" success, "knocking on the doorway to cure"[34]. Non-Hodgkin lymphoma is under consideration for idiotypic approaches in multiple clinical centers worldwide[35]. Promising results are reported for other cancers, too, including multiple myeloma, colon cancer, and small cell lung carcinoma[36]. For some tumors, clinical trials are based on anti-idiotypic antibodies that mimic tumor antigens, including CEA (antibody 3H1) or a particular ganglioside, GD3 (antibody 1E10), both target antigens for anti-tumor immunity[36]. Multiple additional preclinical projects and clinical trials seem to indicate unbroken optimism and continued excitement with idiotypic approaches in the cancer field.

Another group of clinical conditions for which idiotype-specific manipulations had seemed promising are the autoimmunities. Diseases such as rheumatoid arthritis, systemic lupus erythematosus, type I diabetes, pemphigus, myasthenia gravis, and many others, are associated with autoreactive antibodies or T cells or both, inviting speculations on a possible role of a disturbed network regulation in their pathogenesis. Consequently, various forms of anti-idiotypic intervention were considered and tried[37]. A popular therapeutic approach was active immunization with autoimmune T cells, with the aim to induce anti-idiotypic immunity against them[38]. While the interest in idiotypic approaches to autoimmunities peaked before 1995, original papers as well as review articles continue to appear until to date[39,40]. Although some of the preclinical studies yielded promising results, none of the approaches has so far made it to the stage of human clinical trials. In recent years, the scientific literature on experimental models of autoimmunity is dominated by reports on regulatory T cells[41].

Other than at the bench, in clinical studies of autoimmune conditions regulatory T cells receive only little attention compared to another therapeutic approach: Many of these diseases have been shown to improve by intravenous application of normal human immunoglobulins, pooled from large numbers of blood donors. Intravenous immunoglobulin therapy, in short termed IVIg, was originally employed to treat various forms of immunodeficiency. In 1981 a group in Switzerland reported that children with idiopathic thrombocytopenic purpura, a blood clotting defect in which platelets are destroyed by autoantibodies, get better when treated with high dose IVIg[42]. After this had been confirmed[43], beneficial effects of IVIg became apparent for additional autoimmune diseases and presently this therapy is in common use for even broader purposes, including hematologic, neuroimmunological, rheumatic, dermatological, chronic inflammatory disorders and transplant rejection. Recent reviews list up to 35 diseases of entirely different nature which respond to high dose IVIg[44]. The original paper (Imbach et al[42]) has been cited 869 times until today, and more than 2,500 papers on IVIg have since appeared.

The mechanisms of action of IVIg suggested in the initial reports included an inhibition of Fc receptor-mediated clearance of autoantibody-coated platelets, reduction of autoantibody synthesis by 7S inhibition, clearance of unknown viruses, and various other forms of protection of the platelets from the autoantibodies. Michel Kazatchkine and coworkers were the first to consider the possibility of idiotypic interactions of the autoantibodies with antibodies in the pooled immunoglobulin. Indeed, experimental evidence could be obtained that the pooled immunoglobulin contained antibodies that neutralized autoantibodies by what appeared to resemble idiotypic interaction[45]. As a result, network enthusiasts, including Varela, interpreted the therapeutic effects of IVIg as resulting from the reconstitution of a healthy idiotypic network in patients whose network had been disrupted[46]. However, subsequently a host of additional effects of IVIg has been observed, affecting cytokine levels, complement activity, cellular proliferation, cellular cytotoxicity, lymphocyte apoptosis, Fc receptor signaling, etc.[44]. Indeed, Fc receptor targeting has recently been proposed as the major if not the only basis of the beneficial effects of IVIg[47]. The present agreement thus seems to be that network reconstitution is at best one of multiple modes of action of IVIg immunotherapy. For a basic scientist who longs for clear causalities this clearly is a frustrating situation, but in view of the multiple diseases which benefit from IVIg treatment perhaps not a surprising one. In any case, IVIg continues to be a field of active clinical research, in part governed by the network paradigm.

Chapter 14

Hindsight

Those who know the end of the story can never know what it was like at the beginning.

C.V. Wedgwood, British historian

This chapter contains a collection of interviews with immunologist colleagues who have, like myself, been involved in or exposed to the idiotypic network theory (INT) during a significant part of their active scientific career. Some of the key individuals in idiotype research have passed away, but those still living have been approached and many of them agreed to participate. In order to maintain a comparable structure, interviewees were sent a list of 15 questions well before the interview:

Q1.   What was your attitude to the INT in the beginning, say 1975-78?
Q2.   Have you been involved in idiotypic research prior to INT? If yes, what particular aspect?
Q3.   Did you get involved in idiotypic research as a result of INT? If yes, what particular aspect?
Q4.   Did you reject INT then? If yes, why?
Q5.   What are your views now about the former controversy about the Ig/non-Ig nature of the TCR?
Q6.   What do you think now about the former research on suppressor T cells?
Q7.   What about the inducer/transducer/effector concepts on helper/suppressor T cell circuits?
Q8.   What do you think about the former results on soluble helper and suppressor factors?
Q9.   Do you think INT had a stimulating/inhibitory influence in immunological research?
Q10. Do you think that some of the research on INT was useful or a waste of time?

Q11. Do you think that there was a difference in acceptance of INT in Europe vs. US?
Q12. Why do you think the INT disappeared from mainstream immunology?
Q13. What do you think now, do you think the INT was correct, wrong, or to some extent correct?
Q14. What were the paradigms in immunology that superseded INT?
Q15. There are still attempts at idiotypic approaches in clinical research, i.e. autoimmunity, cancer, vaccines. What are your views?

Most interviews were carried out person to person or by telephone. The oral responses were taped, transcribed, and subsequently slightly edited by the interviewees. Some questions were skipped, and some answers pertained to several questions and are therefore combined. Additional questions asked during the interviews are given in italics, but have been kept to a minimum, as were other interjections by myself. Some interviewees responded in writing, and some of those were asked for additional clarification by telephone.

Constantin Bona, person to person, March 23, 2007.
Abbey Aldrich, Rockefeller University, New York.

Q1–Q4
I was very excited in 1975–1978, particularly when, based on the A48 idiotype, I realized that idiotype network is not open-ended.

KE: Could you tell me about the experiment, I cannot fully remember.
We took A48, which is levan-binding, made anti-idiotype, and then Ab3, which has the same structure (as A48). Then we made Ab4 and showed that it had the same structure as Ab2, it means that it binds Ab1[1]. From an intellectual point of view it was not reasonable to believe that the system was open ended, as up to 10^9 number of clones would have to be involved. I also mean the experiments that showed that antiidiotypes have regulatory function, i.e. they can induce suppression, or can induce activation.

KE: Have you done experiments of that type?
Yes, A48 was a silent idiotype. With anti-A48 injected into newborn the A48 idiotype was expressed, it was very exciting[2].

Q6, Q7
Well, in the concept of suppression as originated from Gershon, there is now a big boom. While it was originally CD8 it is now CD4 but the function is suppression. I don't know how long this boom will take, for example the

Hindsight

fantastic marker FoxP3, there is evidence now that it is a general activation marker.

Q8

The factors disappeared, it was clearly shown that regulatory cells need contact with their target cells. The demonstration of these factors involved very crude preparations, and very crude experiments. But there are suppressor cytokines. For example TGF-β is the most powerful suppressor cytokine. Probably there was a lot of truth in the old experiments, only the methodology did not reliably distinguish between specific and non-specific suppression. Now with the cytokine methodology developed we know better.

KE: There were several groups reporting on antigen or idiotype specific factors, including Benacerraf, Michael Taussig, Martin Dorf, Tomio Tada.
There was also Don Capra, who retracted the paper on a helper factor with ARS idiotype. But I think there was some truth in helper and suppressor factors, only the methodology was insufficient.

Q9, Q10

It was very exciting from an intellectual point of view. It showed that the immune system, rather than being a collection of independent clones waiting for the antigen, has connectivity from inside.

KE: You think this is true, it is not disproved?
Yes, it has been demonstrated experimentally. But the methodology used had exhausted the human minds that had been involved, and exhausted the subject. Once this was established it did not go any further, connections were not made to the cytokine network which came afterwards.

Q11

The majority of the data were from Europe and one of the reasons of the failure of the network theory was that in America it was not considered as something important. Bill Paul, he worked on many subjects with many people at NIH, including with myself on idiotypes. After '79 he dropped all of these cooperations and concentrated on one subject, IL-4. Network enemies were Mel Cohn and Martin Weigert. The majority of Americans was reluctant, except perhaps for some of European origin, like Benacerraf who did some work, or somebody like Kunkel who as discoverer of idiotypes thought he might get the Nobel Prize and therefore supported it. But the majority thought it was something unimportant.

KE: You said that American science is reluctant to accept concepts from abroad. Are there any other examples?
Some concepts were so strong that they were impossible to reject, such as

the rearrangement of gene fragments shown by Tonegawa, who was not seen as a true European as well. This is a difficult question.

Q12
The problem is that there are notions which are very fashionable but the fashions do not last a long time. This was also the case for the network. In my mind the major cause that generates them is the grant system. You should always propose something which is new. Even if you only change the name or you only add something. For example, innate immunity. We know from Metchnikow there is non-specific immunity, everybody knew this. They change the label, they found these receptors which they think are important in recognition of something for example in gram-negative bacteria, very crossreactive. They propose that this is now important for specific recognition. I believe it, but it is non-specific immunity which we know from Metchnikow. They only call it now innate immunity and so they have a connection between innate and adaptive immunity. We know this for a long time, as well as that there are antigen-presenting cells. Two or three years ago there was a big boom on antigen-presenting cells, and this also goes on because now there are Toll receptors. These booms in general do not last for very long. Just think of the work on immune response genes many years ago at NIH. Now nobody speaks about immune response genes, they are not in the textbooks any more. Like idiotype network is no longer even in my textbook.

Q13
The attraction of the network was that it is interconnected. In contrast to the clonal selection theory where the cells are preprogrammed genetically for a given specificity and just sit there and wait to see the antigen. Now we knew from Miller's work and Claman's work that T cells and B cells spoke one to another, and here the network failed to find out how they spoke to each other. Particularly as we know that T cells recognize peptides and so on, there are no data to connect this with the recognition of idiotopes by T cells.

KE: There was the notion that T cells recognize idiotypes of B cells.
Yes, but on a very superficial level. With the context of the knowledge on antigen-presenting cells and the presentation of peptides, it was not clarified how T cells recognized idiotopes, it was not shown if idiotopes are epitopes recognized in association with MHC.

KE: Do you know Sercarz' work on the recognition of TCR-derived peptides by regulatory T cells?
Yes. There are now a few studies 20 years later, at the time this was not continued. The concept of internal image was very attractive, and there were many commercial attempts by companies to apply the concept, for

example for prophylactic vaccines. It failed because idiotypic vaccines did not induce memory, which is the main condition for a vaccine. It was tried with hundreds of anti-idiotypes, they induce IgM and short term protection but no memory, even in the form of conjugates that activate T cell help. The companies put a lot of money in this but it failed. A different approach was the idiotype vaccines for lymphoma. But here it turned out that antibodies to common B cell antigens, such as BCL (CD20), were more effective, and they could be used for all patients, whereas anti-idiotypes would have to be made for each patient individually. All of this was disappointing from an intellectual point of view as the internal image really does exist, even at the molecular level as shown by crystallization experiments by Poljak[3].

Q14

What is important is that the clonal and network theories have been developed before the cytokine network came up, and the two were not connected. The cytokines represent another type of network, less specific but very powerful, it explained the connectivity in the immune system, and it did so better than the network. It also allowed to study the molecular aspects of signal transduction, something that the network never touched on, namely what happens between the receptor on the membrane and the activation of the genes, transcription factors etc. There was no connection between these molecular processes and the network, and that was one of the reasons for its failure.

The cytokine network broadened the views on connectivity, it enlarged our understanding of connectivity, including the connectivity with other organ systems. For example IFN-γ has pronounced effects on fibroblasts, suppression of collagen synthesis, whereas on macrophages is has activating effects. The idiotype network was not only very specific to the immune system, it also was restricted to given clones, and did not even include all clones that recognize the antigen. For example the A5A idiotype was restricted to a given clone but did not include all clones reactive with the carbohydrate. It also was restricted to a given mouse strain. The idiotypic network connects T and B cells, whereas the cytokine system connects the cells of the immune system with one another and with other cells not belonging to the immune system, connectivity at the level of the body.

Q15

Particularly in cancer. Kazatchkine made efforts to demonstrate anti-idiotypes in the IVIg treatment, but the disadvantage was that these are big pools, the study could not really prove the origin of the donor in the pool that contributed the anti-idiotype. The use of anti-idiotype in colon cancer disappeared, you know, Koprowski and others. In melanoma, Ferrone uses anti-idiotypes against antibodies to tyrosinase, and claims they shrink. But there is not very much activity now, not even in cancer.

138   Chapter 14

Pierre André Cazenave, person to person, March 10, 2008.
Pasteur Institute, Paris

Q1
I was part of it at the beginning. I think that my attitude was positive, of course.

Q2
Yes, I was involved in idiotypic research prior to the idiotypic network. Especially I worked with Jacques Oudin on the idiotypy of anti-protein antibodies. I had not been involved in the anti-Salmonella work which was the beginning of idiotypy with Oudin. In our early idiotypic experiments we have shown that antibodies against the protein and immunoglobulins without detectable function against the protein, ovalbumin and thereafter fibrinogen, had idiotypic crossreactivity[4].

Q3
Yes, I continued to work on idiotypes after the idiotypic network theory came out, especially working on internal image. I published a paper with Niels Jerne, the work showing there are two kinds of anti-idiotypic antibodies, one set able to recognize the idiotypes, and another set of anti-idiotypic antibodies which in fact were recognized by the idiotype[5]. These are the internal images and it's possible to induce an antibody response against the nominal antigen by immunizing individuals with this second kind of anti-idiotypic antibodies. In this work we have used anti-allotypic antibodies, – it was the b4 allotype (of rabbit κ light chain) – and we were surprised to find that the anti-idiotypic antibodies obtained against the anti-b4 antibodies were able to react with the anti-b4 of all rabbits tested and even with chicken or goat anti-b4.

KE: You probably got to know Niels Jerne pretty well?
In fact, I met Niels before the idiotypic network theory, because he was close to Jacques Oudin. Niels had great respect for Jacques Oudin. My relationship with Niels was more and more close, you know, he spent one year in Paris, and during this time we wrote this paper on internal image. After he retired in the South of France I often visited him because, in fact, his house in the South was only 10 km from a country house that I own. So I have seen him until the end. At the end he was really sick and the last time I have seen him was one week before he died. But he always was fully alert. He had first a cancer of the mouth, I don't remember if it was the tongue or other, and that was cured. After that he had a second independent cancer of the lung, which had probably been latent and then stimulated by the irradiation (-therapy of the mouth cancer).

KE: Do you remember the meeting at the Pasteur Institute in 1973? This was

Hindsight 139

the first time I heard Jerne present the network theory. Did you know about it earlier?
Yes, I remember the meeting, but it was in 1974. But no, to my knowledge he talked about the network before that meeting only with Jacques Oudin, and only several months before the meeting, and not in much detail.

Q4
I don't reject the network theory even now. I can say that it is not enough to explain any phenomenon, but I cannot reject the network theory as a whole. I agree with the point of view that there are instances in which network regulation takes place, but it is not the major regulatory mechanism in the immune system.

Q5
I think that this controversy was due to the techniques we used at that time. The controversy of the nature of the TCR arose from results obtained with antibodies, especially with polyclonal antibodies. Now this problem is solved. At the origin it was because some positive reactions were found with antibodies that were not so specific.

KE: Niels frequently made the point that we should be reluctant to accept evidence on a non-immunoglobulin receptor on T cells. Do you think this played a role?
I myself have never heard Niels say that.

Q6
Concerning suppressor T cells, now people are working on suppressor T cells and name them Treg. With reference to the work of Gershon, you remember that the first work on suppressor T cells was done by the Herzenbergs, suppressor T cells responsible of allotype suppression, and it was work very well done[6]. They never tried to find some receptor, you remember, but afterwards there was a lot of work on suppressor T cells, for example that there was a special region in the MHC, the J region, and so on, and also work showing that there was a special receptor constitutive for part of the sequence encoded by the I-J gene, and another part of this receptor with idiotypic determinants. I mean in particular work of Benacerraf, and the Japanese, Tada. So again, this was probably due to the poverty of reagents. But again, the first observation of suppressor T cells, without any of these problems, was by the Herzenbergs, and this was reproduced by Weiler in Konstanz.

Q7
First of all, this work was done by very prestigious laboratories, Benacerraf, Tada, and so on.

140 Chapter 14

KE: I would like to have your opinion on how so many wrong results could have been obtained.
It was not the time of good reagents and good technology.

KE: This could not have been the only reason for all these wrong results. My opinion is that the network concept was so fascinating to people that they got carried away with these ideas. I don't think that this was scientific fraud, I think it was opinionated interpretation of results.
So you think this was a side effect of the network?

KE: Yes, there were a lot of idiotypic interactions in the suppressor circuits.
Yes, ok. I have never thought about it in this way. This is the first time that I consider the possibility that the network concept was, in part, responsible for the suppressor story. But it is a good idea.

Q8
Again, soluble helper and suppressor factors, it's a problem of reagents.

KE: The first soluble factor that was described was the Munro/Taussig helper factor, and that was clearly fraud.
Taussig, you know, was another problem, exactly.

KE: But then, afterwards, many people described soluble factors, mostly suppressor factors, which also don't exist.
For Taussig it's clear it was a fraud. It was discovered by Ceppellini, because it was a human system, human helper T cells.

Q9, Q10
There is something very positive about the idiotypic network theory. Before, most of the immunologists considered the immune system as a collection of independent clones, totally independent clones, waiting for the antigen, without connection among the clones. And also, it had a role in directing attention to the so-called natural antibodies. Now we know that they are very important. Concerning the natural antibodies, natural auto-antibodies actually, people used to consider that it was a kind of artifact, you remember. And so, probably, the idiotypic network was one of the ideas which were important in considering to study these natural antibodies. And, at the present time, we know that they are very important.

KE: What is their importance?
In terms of history what is very interesting is that, for instance, Rolf (Zinkernagel) said before it's bullshit. And now, he considers, and he published this, that they are very important, for instance in the response, the immunophysiopathology, of viral diseases. And there are more and more publications on the importance of these natural antibodies which, in fact,

Hindsight 141

are now considered as part of innate immunity. They are very important for the regulation of the system.

KE: Do you think that the network theory had a role in recognizing the importance of the natural antibodies?
This is only one factor, it is not the most important factor.

Q11
I think so. Most people working in idiotypic network were Europeans.

KE: If you look at the publications, there was also lots of stuff from the US.
That's interesting, I never looked at the origin of the publications, Europe vs. US.

Q12
One reason is that the idiotypic network theory doesn't explain the regulation. Secondly, the work done on idiotype network, in fact, had arrived at the end. If we consider the different studies, at a given time they were at the end, because they became repetitive, there was no progress.

Q13
It is difficult to give a response, because if it is meant in terms of what was observed, it is correct, correct in terms of the results.

KE: Not everything was correct.
Yes, it is clear that there are interactions between elements, antibodies, inside the immune system. But again, one can say that it is not possible to explain the regulation.

Q14
The present paradigm, Treg. But I consider that a very simple view, to think that there is only one kind of Treg, CD4, CD25, plus FoxP3. This is an over-simplification of the system. There is some evidence that there are other Treg or suppressor T cells. If this is a mechanism which is important for regulation there must be, of course, several kinds of Treg. If this is so important, during evolution it has to be duplicated to keep several kinds of cells able to regulate.

Q15
Firstly, I am not sure that it is the best way to do vaccines and secondly, when they use anti-idiotypic antibodies of internal image type it could also be dangerous, because you could stimulate other antibodies than that directed against the infectious agent, and you have the risk that some of these antibodies are dangerous, for instance auto-antibodies or antibodies able to stimulate autoantibodies. In autoimmunity, do you mean IVIg? I

think it is a very simple view of autoimmunity. In severe autoimmune syndrome there are a lot of auto-antibodies directed against many self-antigens, not only one. It is clear in lupus, even in diabetes. So I don't know. With respect to cancer, if you have cases in which this kind of immunization is effective, it is very difficult to apply. It is only for people able to pay a lot of money.

KE: Is there anything else you like to say, in terms of hindsight or retrospective views?
Independently of the idiotypic network, I think it was a time of discussion between people, those working on idiotypic network, but also with people outside of the field or even against the network, just think of Mel Cohn. I think these were very interesting interactions among scientists. This is no more the case.

KE: The network theory was the last comprehensive theory of the immune system, as far as I can tell. There is no successor.
At the present time we see a very reductionist approach. Most people are working on signaling, this is interesting, you have to know the signal chains, but you cannot explain how the immune system works. Signal chains are like the Krebs cycle, if you know one more or one less element is not very important. Signaling is important for cell biology but not for how the immune system works as a system, and a system connected with others, which is one of the most important fields in immunology at the present time. I am talking about basic immunology, so in terms of people and money spent for research, signaling and Treg are most prominent. There is now a repetition of results on Treg. There is nobody now who has a larger view on regulation. This is the case in basic immunology, and even more so if we move to immunophysiopathology.

Antonio Coutinho, by telephone, March 27, 2008.

Q1
My attitude to the INT in the beginning was, of course, very positive, very excited, for various reasons from my short experience before. As you might recall I was in Göran Möller's lab, where I was exposed first to the theory, and then coming to Basel where I was more exposed to it. In Möller's lab, as you might recall, we had come to the conclusion that B cell activation required what we then called mitogen receptors, called now TLRs, and we sort of radicalized the theory in the sense that immunoglobulin receptors would serve little or nothing in terms of activating the B cell. The INT was excellent in this context because the assumption in our theory was that B cell activation by non-immunoglobulin receptors would solve the problem of self-nonself discrimination at the B cell level such that self-nonself dis-

Hindsight

crimination would not be determined by variable region repertoires. The problem was if you eliminate all autoreactive cells you would end up with no repertoire, because of the degeneracy of antigen-antibody interactions. Therefore the INT fitted very well in that way of looking at things, that is there is no self-nonself discrimination on the variable region basis, and the existence of the network was never a problem for me, in so far as it didn't impinge on problems of self-nonself discrimination.

KE: If I remember correctly, Göran Möller was not a fan of the network theory. He was not a fan of the network theory, as you say, I guess he did never pay too much attention to it. But some attention he paid, because when I came to the lab he told me that there are a number of people in the field that one should listen to, and Jerne was one of them. When I came I had no experience in science, I was an MD, and I was overwhelmed by the many papers that were coming out. So he had these two lists of people, one of people of whom you had to read everything they published, even if very often they would be wrong. The other was a list of names one should not spend time reading, whatever they would publish. Möller was very much black and white, he still is. A radical so to say.

The excitement with the network theory was, first, it didn't impinge on self-nonself discrimination, and second, it contributed to what I think medical education has a lot to do with, that one is trained by necessity of the profession to look at things in an organism-centered manner. One of my biggest disappointments in my beginnings in immunology was that it was fragmented in cells and some molecules and we didn't have any way of considering a system that was regulated inside of the body, the whole body. That's what the theory brought, and I think it was excellent in that sense, and I was very positive to it, no doubt. Then, coming to the Basel Institute, I was more exposed and more engaged in discussions and experimentation so this got me even more excited.

Q2
No, I started immunology in 1972, so I couldn't have been.

Q3
Yes. What I did, as you know in part with you[7], I learned a lot of network from you, idiotypic work at least. What I started trying to do in Basel was actually determined by the theory itself. Not only the quantitation of idiotype precursors and idiotypic interactions, of phenomenology that you and others had described that we confirmed using limiting dilution as you remember, seeing these phantastic things that by priming with either the antigenic ligand or with the idiotypic ligand, you get expansion of the other specificity as well, you remember. All of that implied that there was a regulation that applied to the whole variable region, how they interact with each other, and how the system is regulated.

What I then did in Basel determined a good part of the next years, what we called then the internal activity of the system. That is, if the theory was correct, this in some way should have functional consequences for the cells, the system should not be at rest if it's let alone. There should be some internal autonomous activity inside the system. The only paper I have together with Niels and Georges Köhler, signed first by Luciana Forni[8], that injecting IgM into normal mice, monoclonal or polyclonal IgM, you get more of the same specificity. So it's what then Jacques Urbain started to call the hypercycle, that you get more of the same specificity, and it was still based on this internal activity, thinking profoundly influenced by Jerne. To me it meant that we have to understand natural antibody production, which is the activity of the system at rest. This has a lot to do with views on self-nonself discrimination that developed some years later, especially under the influence of Stratis Avrameas, his demonstration of a lot of autoreactive antibodies in the natural antibodies, what we started calling the physiology of autoreactivity. This included largely these idiotypic notions.

What became clear in a way was the intimate association between what is idiotypically connected, so to say, and what is autoreactive. This came about for two reasons, first because, still in Umeå after Basel, we had produced these large collections of monoclonal IgM antibodies from newborn mice, fused from the spleen as they are born, and then we looked at their reactivities, and they were essentially all autoreactive, and a very large proportion of those reacted with each other. A piece of results that then John Kearney reproduced, that the repertoire of antibodies starts as a densely connected network, somewhere between 20% and 30% of all possibilities are positive, at least by the *in vitro* tests one can do. What John Kearney did was to use these early interactions to study how they shape the future repertoire, a very important piece of work. We rather looked at them in the context of the natural antibody production in the adult, the ones that are retained in the adult. So we produced reagents, a number of newborn monoclonal IgMs which every newborn produced, and then we followed them in the adult, and we found that the natural production of those molecules is extremely sensitive to administration of either the same molecule itself or to anti-idiotypic antibody against them. You can give something like 10 ng of an IgM monoclonal and the mouse that is producing it at given levels in the serum will remember for the next three to four months that it had got 10 ng of it some time ago. So it's an extraordinarily sensitive system.

All of this was somewhat reinforced by the fact that looking at the genes of these connected antibodies of the newborn, as done by Dan Holmberg and John Kearney, it turned out that many of them were encoded by these D-proximal VH genes, for which other people had shown that in the developmental program they were expressed first. In addition, most of these connected antibodies of the newborn have no TdT residues, and as you know, expression of TdT is also developmentally controlled. So everything was in place to conclude that indeed there is a genetic developmental program to start with sticky antibodies. The first antibodies to be produced, in over-whelming majority, are sticky. They stick to each other, and to many other proteins in the body. Ok, that's clear, and I think it was never disproven in any way.

Of course, the network theory has many implications, in clonal respons-es and whatever, in regulation and many other things, but I thought, and John Kearney too, that this was one way of testing the theory: Is there a net-work or not when the system starts? And the answer to that is definitely yes. But again, as I was saying before, very much again in the context of autore-activity, the same sticky antibodies that react with each other, react with lots of other proteins in the body, not restricted to idiotypes.

Then, in Paris, we came back to this problem, which we never left so to say: What is the system at rest? We started separating, in adult mice now, resting cells from activated cells and looked at their repertoires in various ways, by limiting dilution, by hybridizations, etc., and the conclusion was rather straightforward, too, and again was never proven wrong. In the adult now, essentially all autoreactive cells and cells that show connections with other antibodies, are activated cells, have markers of activation, can be depleted by anti-mitotic treatment, can be separated from the small cells in gradients, they have all the characteristics of activated cells. And essential-ly all the activities that you can measure in limiting dilution against exter-nal antigens are, of course, resting cells. It made a lot of sense, what is being interactive internally is activated. It's not pathogenic for reasons that we can discuss later.

The radical test of that was to look at this type of cells, immunoglobulin production by activated B cells, also T cells, CD4s, CD8s, and so forth, in antigen-free mice. The Burnettian theory says, number one, there is no autoreactivity, number two, the system is at rest before stimulation by exter-nal antigen. So just take an animal that has never seen an antigen, and you might remember that these animals have a crack in the spleen, they are so germ-free and antigen-free that they have no lymph nodes, no intestinal lymphocytes, no Peyer's patches, no nothing, but the spleen is normal. Spleen, bone marrow, and thymus are of absolutely normal size. And the spleen contains just as many activated T and B cells as a normal mouse, lev-els of IgM in the serum are actually higher than normal, since they don't make IgG or IgA at all, they have IgM slightly above normal. The conclu-sion was that, somehow, there is enough autoreactivity or idiotypic connec-tion to activate these cells, and the true physiology of autoreactivity is in that set of cells[9].

This type of interest was developed very much with Francesco Varela, who was a specialist in networks, mostly in neural networks but then he got interested in this at some point. We started developing models to deal with this which ended up in what we called "second generation immune net-works"[10]. We came to say something that nobody liked, neither the idiotypic network people nor the ones that were against it. Because what we proposed is that, indeed, there is one part of the system that is entirely Burnettian, disconnected, evading productive interactions with either self or idiotypic network, and that we called the peripheral system. It's made of small cells and, at least functionally they are outside the network, the network doesn't contribute anything to their physiology. They are resting cells, just the same as they are produced in the thymus and the bone marrow. By definition, they are not functionally impinged upon by the network or by autoreactivity. But, side by side with this there obviously is another compartment of the system, that we called the central immune system, that involves about 10% of the cells, T and B cells, which are autoreactive and connected. As an addition, this part of the system which is autonomously activated and connected is predominant in embryonic and neonatal life, in the first week of life it's 60–70% of the cells, and only when the mouse gets past 6–8 weeks, in the adult, the autoreactive system comes down to 10% or so.

Niels didn't like this at all. He was in the South of France and I was in Paris, we discussed this a lot. In the end he always told me this must be wrong, you can't have two immune systems, and so on. So we spent a lot of time discussing how could it be that the activity of the network could also impinge on the small cell system, which would be easier, one can find many ways that we never tested, but Oudin's old experiments[4] and the reactivities that we saw together in the limiting dilutions[7] already suggested ways how this may well be.

Then the modeling part was trying to be more realistic. If you consider what many people have discussed before in the network context, a variable region made by rearrangement in the bone marrow by a particular cell that comes out, this is only 10^5 molecules localized to a specific cell, so this will impinge very little on anything else. If you consider what is in the serum, and this is again influenced by Niels, there are orders of magnitude more variable regions free in the serum as compared to what is attached to cells. Therefore, in these modellings with Francesco Varela and John Stewart, this was considered and we separated these two things, and connected it to the previous theory on peripheral and central immune systems, autoreactive and Burnettian. In other words, my conclusion is today, after all of these years, at least part of this Paris thinking, that is that autoreactivity is physiological and necessary to assure tolerance, part of that is fully vindicated today, with the story of the regulatory cells selected to be autoreactive in the thymus, and all that. Looking back I still think this is roughly correct that there is indeed a Burnettian part that takes care of clonal responses to non-self, and there is the Jernian part that has a lot to do with the relationship of the immune system to the body, and insures tolerance on the basis of positive selection for autoreactivity and for connection. That's my current philosophy.

Q5, Q6
I think these have a lot to do with the fact that most of these experiments,

because the historical time was favorable to it, were done in tissue cultures in conditions that were not quantitative. This may be the major reason for all of this. And the accessory methods that were not ideal. Of course this doesn't apply to all what was done on the Ig/non-Ig of the TCR or on suppressor T cells, but in suppressor cells it may apply to almost all of the work. And there was also that level of maturity in the community that was expecting to see something like that. In a way the same thing happens now with the regulatory cells, because it was clear that the deletion model of tolerance could not be fully correct.

Q7

I am amazed all the time how many new cell types there are. If you classify cells, let's say, you have 25 cytokines, then you can organize classification on a 25-dimensional space, and you can probably identify many hundreds of cell types. The question is if this is a commitment process, a developmental process that commits the cells irreversibly, or not. And I think in most cases probably not, maybe in some cases yes. We are far from seeing the end of the road in those aspects.

Q9

It could not have had much negative influence. Not so many people were involved in doing work in network theory. In proportion, many more theoreticians than experimentalists were involved in the network theory. I think it had positive consequences for the development of theories on networks in general. I recently was in a meeting on networks, they had asked me to come, to my surprise, because I was not a theoretician like Francisco Varela or John Stewart or so, but they asked me to come there because many of the things they had read about the idiotypic network modeling were now very much elated and popular in the context of these scale-free networks, of these more-than-small-world networks as they are called. Modern modelings and network theory on the small-world networks and scale-free networks were very much used and developed in the beginning within the idiotypic network theory context. That is a contribution that often one doesn't appreciate, certainly a positive one. And yes, I could say that perhaps some of the people that ended up doing the early experiments on regulatory cells, I am thinking of Don Mason, ourselves, Sakaguchi, LeDouarin, Herman Waldman, but not so many more, their essential reason to start these experiments was the conviction that autoreactivity was physiologic, and this is very much related to the idiotypic network theory itself. So for myself it had positive consequences, and I do not see many negative ones in immunological research because there were not so many people doing it.

For my colleagues and myself it was useful at least in one respect, it forced us to think in terms of organism-centered biology, global properties, and if I want to make a criticism to the original network theory of Jerne, it is that it was not organism-centered, it was immune system-centered, so to

148   Chapter 14

say. It isolated it from all the rest of the body and I think that was a major problem. For us, it helped a lot to think in organism-centered terms, in what is now called systemic biology.

Q11, Q12

Europe *versus* US is related to the mainstream. I think the things that remained in mainstream are the things that sooner or later solve problems, and the idiotypic network theory so far has created more problems than it solved. The American attitude to life, if one can say that, is very pragmatic. From a very utilitarian point of view the network theory has been rather useless so far. Because it is a systemic problem, you can't clone it, you can't put it in data bases, and therefore it is not appropriate for the modern agenda of biology. For the last 15 years the agenda is component analysis, with some positive consequences, for sure, but also with some negative ones. I used to tell my medical students that we know every single little molecule, cells, and genes, that participate in allergic reactions, and we continue to treat allergies like we did 60 years ago or more. Because we don't understand the whole thing together. And this is the direct result of component analysis. Because this is the dominant view in modern biology, the network theory could not survive.

Q13

I think it was to some extent, a large extent, correct.

Q14

I really don't see any. Even the modern views of dominant tolerance had been already there. One of the original quotes in Jerne's first paper was that the function of society is to suppress the individual. So there is no paradigm that has superseded the network.

KE: How about the cytokine paradigm, or Treg. But you may not see these as paradigms.

No. The Treg paradigm is perhaps a paradigm, in the sense that it embodies the same basic thought, that society suppresses the individual, and we know today that societies organize themselves so as to have some of these guys, so that's it. The cytokines, yes, but they have a problem, that I'm sure you share with me, they may regulate everything but they have nothing to do with the specificity of the recognition of the effector function. Again quoting Jerne, who used to say that many things are required for the grass to grow, but they don't determine the kind of grass. Of course cytokines are very important, but they don't determine what we in immunology are interested in, which is the specificity of recognition.

Q15

Why not? On the other hand there are many ways of manipulating variable

Hindsight 149

regions and clones. In these attempts I don't see the difference between manipulating things with paratopes or idiotopes. In the past I have gone a little bit further, in the sense that one of the reasons why the network theory failed was that in most cases people used the network theory to study clonal responses, or to manipulate clonal response. In a way, this is to deny the theory itself. To use idiotypes as clonal markers is fine, but to use them to infer properties that pertain to the Burnettian part of the system, that may be wrong.

Ron Germain, by telephone, Feb. 7, 2008.

Q1, Q2, Q3
Looking at your questions is bringing back all sorts of memories. I was obviously very intrigued by it (the INT), because of the work we had been doing with suppressor cells and suppressor circuits[11]. We obviously became very interested in idiotypes as an aspect of immune behavior. We were finding that the factors and the restrictions between the cells were IgH-linked and that anti-idiotypic antibody seemed to matter for some of the cells[12], so in those experiments we seemed to be finding results that fit very nicely with at least a portion of the model. I actually just wrote something for the 50th anniversary of the journal *Immunology*, a historical piece on what was going on in the early days with suppressor cells, what potentially went wrong and how this all relates to current things (with Treg)[13]. You can find it all in the *Immunology* piece.

Prior to the elaboration of the INT, no, I don't think I was thinking along these lines. It was in the midst of it becoming an interesting model that I started to incorporate these ideas into my experiments. I became interested because there was ongoing work in neighboring labs studying idiotypic networks and providing reagents that we could use to probe the systems we were studying, and it seemed appropriate to do that. I would have to think hard to say how much of that decision was opportunity and how much was intellectual – like "Yes, this might be the real answer to how it works." I am not sure I can separate these two parts.

Q4
No, I didn't. Again it is always hard in hindsight, after this length of time, to say, did I buy into it as being everything it was claimed to be at the time, or did I have reservations where I wasn't sure was it the fundamental structure of the system as opposed to occurring in the system. This difference – it is a little bit of immunophilosophy. It is the difference between what does happen and what has evolved to happen. I can give you an example I always use, from Al Singer's studies. When Ada Kruisbeek was in Al's lab they carefully documented that there are CD8 T cells that kill based on recognizing class II, not class I. They showed, however, the selection of those cells

depends on class I and not on class II in the thymus. So the cells clearly exist, they are selected as if they were conventional CD8s, but they look at allogeneic class II. There is a reality there and you can look at it in two different ways. On one hand, you can derive your entire intellectual construct about how the thymus works based on ensuring that these particular T cells are made, as an important part of how the system evolved. The alternative is that the system evolved for its own purposes that are different from making those particular cells, but the nature of how the thymus works, the nature of recognition, of degenerate TCR specificity, a lack of negative selection for allorecognition, is such that you don't eliminate those cells because physiologically they are irrelevant in the individual, unless you are a surgeon doing a graft. That is, it may be medically important to know that those cells arise, they definitely are there, but you shouldn't use those cells to construct your model as if evolution operates to make sure the thymus made these cells. I don't know when this transition ('should happen' to 'can happen') occurred in my own mind with the idiotypic network. The system has enough diversity that if you confront it with a high enough concentration of a certain ligand under conditions in which you can break tolerance, you can show that you can get a response. The fact that you can make anti-idiotypic responses, and once you made them they can influence the system, doesn't necessarily mean that the system has evolved to use those types of reactions to drive its operation. But you can influence its operation. So, from the perspective of "Could this matter in certain cases, just like these allo-class II specific CD8 cells would matter in a graft situation?" the answer is yes, you still would need to think about it. But it would be different from a theory or model saying that the immune system has specifically evolved to have this idiotypic network reactivity that controls its function. The question is, is it part of the fundamentals of the system or not? This transition in my thinking occurred after I started working on idiotypic properties of the system. I can't tell you when, but much earlier than today. It wasn't in my mind when we were first generating the data, however. I want to be honest with my own thinking.

Q5
The simple answer is that it is obvious by molecular analysis that the TCR is Ig-like and depending on how crossreactive or diverse an antigen you have and what type of cross-species immunization you use you could get crossreactions. That may even explain some of the idiotypic relationships that we saw in the suppressor data we were talking about. The other point is, and that includes the suppressor work I was involved in, some experiments were really good, but some experiments were not so good, and our reagents were not as good as they are now. And so errors were made. I am not going to say: This was an error and that wasn't. There was a limitation of the reagents, which is not bad science but a limitation, but we could also have been more cautious, more concerned about those specificity limita-

tions. And of course, some things were just not that well done. In combination, this all fueled the controversy.

Q6, Q7, Q8

This is all laid out in the *Immunology* piece[13]. Basically, there are many observations made then that still hold up. If you look at the cyclophosphamide sensitivity of suppressor cells, it has been found to apply to Treg; there are all sorts of similar parallels. Do I think that everything that was done (working on suppressors T cells) was right, data in particular as opposed to interpretation? No. How much was just bad science and how much was the limitation of how we did science then, is not so clear either. One of the reasons I say this is, in lots of other fields and other aspects of immunology and other fields of biomedicine, you go back and look and there is all sorts of stuff that is wrong in the literature. People believe they know what the answer is and then things are revealed as poor results or misinterpretation. The real difference here is less whether the work was as good or not as good as in other areas – there was a psychology here. There were a couple of things that turned people's minds, that changed perception here, really independent of the actual science. For example Tregs, they are not crashing and disappearing, even though you can see there are clear errors in some of what's been reported or some of the methods that have been used for studying them. Yet the field (Tregs) seems bulletproof; it is unlikely anybody is suddenly going to deny that these cells exist, although I am sure there are many, many claims that are exaggerated or incorrect. Whereas in the other field (suppressor T cells) there was a sort of undercurrent, almost looking for it to fail, and partially that was because there already was – and that refers to the helper and suppressor factors – evidence that things were made up. You know, the Taussig/Munro factors. With that undercurrent, with molecular biology and monoclonal antibodies, with immunology becoming a more rigorous field than it was in the past, the suppressor area was ripe for being attacked, from a sociological standpoint. So when certain experiments failed or clearly had errors in them, rather than people going back and asking what can we be certain about and what did we miss or misinterpret, it basically became "It's all garbage, let's ignore it." And it had to be rediscovered in another guise. The answer is, there was a rush to judgement there and the baby got thrown out with the bathwater. Many of the intellectual arguments that Dick Gershon made, that you had to have some negative regulation, something that's controlling the effector limb of the system, are correct. And there were aspects of the circuits that were described that have some relationship to current results. If you look carefully you realize that you need IL-2 to activate the function of Treg and to maintain them, and most of that appears to come from other T cells, so these other T cells in that sense are 'suppressor inducers.' If you go to the Asherson/Zembala work, they realized that mediation of suppression was due to unspecific factors at the end of the day. Some of these appeared to work on antigen presenting cells, and this is what IL-10 does, right? Think about the action of Treg at the level of dendritic cells as revealed by the imaging experiments that people have done now, where you don't get effector T cells clustering around the dendritic cells. If you actually look at the data in the immune regulation field in general, many of the things that were true in the early suppressor work, about there being at least an inducer cell and an effector cell, about the effects being mediated at the APC level, about the release of an unspecific mediator through antigen-specific stimulation, all are true. They are not wrong.

The issue of whether CD8 cells were involved, well, it is important to remember that it wasn't CD8, it was Lyt2, based on antisera. There were also cells that were Lyt2/Lyt1 double positive. The studies weren't all done perfectly. But some of the problems were reagent limitations, at least in part. It would have gotten cleaned up if people had continued to do this work with the new tools. So my sense of it is that there were some bad turns in the road, in some cases in the sense of "That's the best one can do.", in other cases where things weren't as good, and yet there was a lot of reality to it. People forget about immune suppressor genes, the Is genes on which I worked with Baruj (Benacerraf), they all mapped to the class II locus, and the Tregs are all based on CD4s, which recognize class II. If you look at all of these things, they resonate when you go back and look. It's the difference between allowing a field to evolve and correct errors, so as to come to the right conclusions, *versus* just saying it's all junk. It would be like saying today that AID had nothing to do with hypermutation and switching because many investigators believe Honjo is wrong about how it works on the molecular level. There are some things that are right and some things that are wrong, and you figure out which are which and then you move on.

Q9

I think it did have a stimulatory effect, but some of the things went in the wrong direction. I think there may not have been enough dialogue along the lines that I mentioned earlier in terms of this (idiotype responses) being something that just can happen vs. being a critical aspect of the organization of the immune system. We could have discussed that a little better earlier on, trying to figure out what the real value of the model was to the field. However, I don't think it inhibited anything, people didn't fail to do important work because this model was out there.

Q10

Yes, some of it is useful. For example, anti-idiotypic responses, responses to the variable regions of engineered antibodies, are critical to consider when you are making antibodies for use in humans. Clearly, anti-idiotypic responses can occur and they can have biological or medical effects. Understanding the fact that one can get such responses is not a waste of time. The question whether energy, resources, dollars, were wasted in pursuing that model (INT) as a model and trying to show the whole system to operate in this way, guide the diversity in antibody production and so on, actually those experiments would have better not been done. But none of us are good enough to know early on, when there is an interesting model out there that we are all testing, what exactly are the right experiments, and there are always excess studies, you only have to go back to the Treg stuff. I recently was discussing the funding system in the US and whether the available money was being used optimally in the grant system, and I said to someone arguing for a higher payline, why don't you go to the poster session in a big meeting and walk down any aisle on a given topic and then tell me that all of those experiments are worth doing. The particular ones these days would be a Th17 aisle or a Treg aisle. The answer is no. There's a lot of stuff there, and most of it is actually not that interesting and probably some of it is even wrong. We still do it. Much of it is a waste, much of it is done because everybody else is doing it. This was true for this (the INT) model, but it doesn't mean it was any different from any other interesting hypothesis where there was enough reality to the biology that people could work on it, which is not always so because of limitations in terms of technology. People could do it in this case, and you got overshoot and then contraction. How wasteful overall it was (work on the INT) I don't know.

Q11
Yes. Absolutely, more in Europe than in the US. But I think for another reason. It wasn't the specific model. It may be changing now, but at least 10 to 15 years ago it was very apparent to me that European scientists were more taken with trying to think about the larger picture of how something is working. Not a theory in the Mel Cohn sense – "I figured it out in my head and it has to be this way; intellectually we know this has to be the answer." – but I mean a model like the network theory or anything else, there is more interest in that in Europe, more of an interest in a 'philosophical' consideration of the operation of a biological system. In the US it is much more practical, many investigators just try to generate a lot of data. I never could imagine generating all the mouse strains and all the results some people have generated. I recall especially one time when I received an exceptional reception in Europe to a talk that conceptualized the question "Why would you want to have positive selection and why would a self-ligand be valuable?", then presented some simple experiments showing that the concept appeared correct; the audience loved it. In the US they wanted to know why we didn't do, oh say, 15 more knockouts and generate tons of more data. There clearly was a difference, but I think it's probably changing a bit now.

Q12
For the reasons I was talking about. Many now think that idiotypic responses are something that can happen or do happen, but no longer are taken with the notion that such reactions are the controlling fact or the dominant

regulation here. That's an intellectual conclusion. The other reason is that although you can get monoclonal anti-idiotypes, a lot of the work at the time was done with polyclonal reagents, and as the field moved to recombinant cytokines and multiparameter flow cytometry and transgenic cells, it became easier to study other aspects of the system and everybody could see whether they got the same answer in their studies. There is also a trend to it. There may be more to the idiotypic network than we accept now. It's a little bit like the old suppressor area, where the field just moved on, it's not the hot thing to think about. I think there was the recognition that it (idiotypic interactions) was a piece, but it wasn't the major piece it was originally proposed to be.

Q13, Q14, Q15
In terms of the paradigms I think it was antigen recognition in its various guises, whether for NKT cells, obviously in B cells, or in T cells in terms of peptide/MHC. You can molecularly define the ligands and show that they definitely control the behavior of the relevant cells. We can show that if we interfere with any of those recognition systems we eliminate an important, reproducibly measurable part of the immune response. This answers the question if something is a necessary feature of the system, whereas it is less obvious that the idiotypic network is a necessary feature of the system. It's like micro-RNAs. A lot of knockouts of micro-RNAs don't have an obvious phenotype right now. Does this make people think micro-RNAs are not important in biology? No. They probably tune things, and it's only under certain conditions that you reveal gross defects. If you knock out a gene you need, you get a lethal phenotype, if you knock out a micro-RNA, you may get a heart valve defect, or you may not see very much yet, and you don't know what to look at. I suspect that that's also true here. Because we now have the means to identify the absolutely critical components, this doesn't mean that other features, idiotype as one example, don't tune the system or adjust its behavior. For the day to day thinking about immune function, it may not be so important, but there may be pathologies where it comes into play. I already mentioned antibodies as therapeutics. Clearly it matters there and it still has a role such as the successes in vaccine treatment of particular lymphoid tumors, although this kind of personalized approach of doing things is always very difficult, on a practical basis, but it can work.

I actually would be very interested in what the other Klaus (Rajewsky) thinks about all this. I certainly do remember your A5A stuff.

KE: An interview with Klaus will also be included. We both agree that experimental errors were made and that data were interpreted in a biased way.
I think for all of us, we were taken with various thoughts. Yes, we didn't have the rigor built into the way we did experiments, ruling out things that could have been observer bias, for example doing everything blind. You may or

Hindsight

may not know this story: as one of his first projects in Baruj's lab Steve Burakoff was to repeat the Taussig/Munro helper factor work. And he failed, utterly, and for good reasons, because it wasn't true. The reason that he did not get sucked in, in spite of prevalent belief around the lab that his experiments should work, was that he coded every plaque slide with a random number code, he never knew what he was counting. I know nobody else who ever did that. I remember the days when he got home and it took him hours to decode all of this, and he came back to the lab frustrated because things were not working. But he was successful because he did everything right, showing that it didn't work when it shouldn't have worked. I think there is a real lesson to learn in this.

Heinz Köhler, written response, and by telephone, August 21, 2007.

Q1
Positive, I published our work on INT in first rate journals (*Journal of Immunology, Journal of Experimental Medicine*). You know I was for a year in Basel, when Niels (Jerne) was there. We had a lot of discussions, but Niels was very excited about Tonegawa's work (on the rearrangement and somatic mutation of antibody genes). He sort of walked away from the network concept. That was sort of disappointing to me.

Q2
Yes. Using anti-Id antibodies to suppress specific immune responses *in vitro* and *in vivo*[14]; discovered that during an immune response auto-anti-Id antibodies were detected in mice[15]. Anti-Id can specifically suppress in neonates a specific immune response for months.

KE: What was your experiment to demonstrate auto-anti-idiotype?
We immunized a large group of mice with, I believe, phosphorylcholine (PC) coupled to KLH, we published this in *PNAS*, I gave you the reference[15], and we determined the titers against PC using hemagglutination. We then coated red cells with the T15 myeloma protein, the dominant idiotype of the anti-PC response in BALB/c mice, and we detected a titer there too. Sheep red cells coated with this idiotype showed agglutination with the sera of multiply immunized BALB/c mice. It was published, I think, in 1974. Two years later Humberto Cosenza, at that time at the Basel Institute, did a hemolytic plaque assay, and detected also plaque responses against sheep red cells coated with T15[16]. In this case it was during a primary response to PC. It confirmed our results.

Q3
No. Our lab was one of the first to use INT experimentally to regulate immune responses and helped to establish the field.

Q4
No.

Q5
The controversy existed because the methods to resolve the issue did not exist at this time. Only after cloning the TCR it cleared and showed that TCR belong to the Ig family. Some of the immunological detection of T-cell Ig could be due to cross-reactivity of anti-Ig reagents with the Ig fold of TCR.

KE: What about the heavy chain allotype linkage of T cell idiotypes, and the sharing of the VH region by T cell receptors?
I have no opinion about that, I don't know these data in detail, I only know Jack Marchaloni's data for which crossreaction is a pretty reasonable explanation. We now have sequence data and do know better.

Q6
The work has been to large extent confirmed by recent work with Treg cells.

Q7
Helper and suppressor cells secrete different cytokines with different biological effects.

KE: How about the results on suppressor inducer and suppressor effector cells as described by Benacerraf, Dorf, Greene, and others?
I remember these results only vaguely. The data probably are ok. Now we understand it better, but it was basically a name change.

Q8
Nothing, since I am not familiar with this type of work then and now. I cannot comment on this.

Q9
Definitely it had great stimulating effects. Idiotypes as specific tumor targets were explored (Ron Levy and others). It led to founding of a major Biotech industry (IDEC). Use of idiotype as specific reagents to monitor immunotherapy with antibodies. Understanding of how IVIg works and how to use IVIg in auto-immune diseases.

KE: Tell me more about the IDEC company?
I was recruited by the founders of IDEC, because at the time I was working on idiotype regulation. They recruited me because Ron Levy had done these clinical studies using customized anti-idiotypes to treat lymphoma patients. That seemed to work, so IDEC was founded to explore that concept further. I helped them to establish the laboratory, I think I was

Hindsight

employee no. 3. Without Ron Levy's work on idiotypes IDEC would not have been founded, I think. It is now one of the top biotec companies, they merged with Biogen.

KE: Did they ever manage to develop anything idiotypic for the market?
No, but we tried. First of all, the customized idiotypic therapy did work but it was not cost efficient. And it took too long. The average time required to make a customized anti-idiotypic antibody was about nine months, and for most patients this was too long, and too expensive. We also did some clinical studies using anti-idiotypes in HIV infection. The idea was to suppress the major restricted antibody response which is not protective because the virus rapidly mutates. If you suppress that, we hoped a non-restricted response would come up that would be effective against mutated virus. We did have some evidence for this in monkeys. But the trial did not lead to any therapeutics.

Q10
Any solid and reproducible research is no waste of time or resources since it increases our knowledge. Even if a given work does not produce a tangible monetary return, it is not wasteful.

Q11
Yes, Europe, in particular the Basel Institute was a hotbed of work on INT.

Q12
INT has failed to produce therapies or approaches to develop therapies most of the time. I can see two reasons. One is that idiotype-based vaccines require strong adjuvants. In mice idiotypic vaccination works, as we have shown as well as others. But it required strong adjuvant, they all had to use complete Freund's adjuvant. For humans there is no FDA approved strong adjuvant, so the trials in humans failed. The second point is that idiotypic approaches in humans have to be very individualized. Each individual has different idiotypes, major idiotypes are the exception. So one has to work with customized vaccines and this is not cost effective, not practical.

Q13
There is no doubt that an immune network exists. However we are still in the dark how to use INT to create cures of diseases.

Q14
The INT has not been replaced with any global theory in immunology.

Q15
INT will continue to dissect and improve the therapeutic effects of IVIg in autoimmunity. Idiotypic vaccines suffer from lack of potent adjuvants for

158   Chapter 14

human use. Idiotypic vaccines and therapy can be improved by a personalized medicine approach based on genetic profiling. Idiotypic vaccines have worked only in inbred or closely related animals.

William (Bill) Paul, by telephone, Feb. 12, 2008.

Q1–Q4
So what are we talking about?

KE: We are talking about idiotypic network!
Oh! I haven't thought about that for a long time. Let me pull up the questions that you sent. Here it is, I got it. I first heard in some detail about Jerne's network ideas from Ben Pernis. Ben was coming to the US then, if you remember. You probably remember this with better precision than I do. It was just at the time when Ben was coming to Columbia (University, New York). He was the first person from whom I heard a description of Niels's. Initially I thought it was very interesting. This concept that there would be a whole series of nodes, and that each node would have components that would recognize other nodes was attractive. And the notion, of course, that an individual receptor would have the ability to recognize both antigen and idiotype, that is that antigen and idiotype would mimic one another seemed like an attractive idea. I was taken with it and I believe, actually, that Constantin Bona and I wrote a little paper together[17] but I never did very much bench research on the network*.

In a sense, you can argue that intellectually there is a lot of similarity between idiotype network and the concept of self-recognition by T cells of peptide-MHC complexes that generates the T cell repertoire. I don't know if you would accept that but for me, thinking of it now at a distance, one could have argued, though I don't think it was intrinsic to the proposal, that repertoire generation in the B cell system could in fact have been very influenced by immunoglobulin itself, just as the MHC sculpts the T cell repertoire. So one could have taken that position. If I had to describe my intellectual journey with idiotypy, it would be something of the notion that idiotypes might determine the nature of the B cell repertoire.

Q5
This is of course a little different. I go back a long time on this stuff. You know when I was with Baruj (Benacerraf) and afterwards I became very interested in the whole question of how the MHC was involved in regulating immune response; indeed, at this time some people thought that the MHC encoded the receptor. I must say I was always on the non-Ig side, in

---

*   W. Paul and C. Bona were coauthors of 11 papers relating to idiotypic regulation between 1978 and 1983.

Hindsight 159

the sense of Marchalonis – who where the other people? – who was strongly pushing the Ig-nature of the TCR.

KE: How about B cell idiotypes on T cells?
I was never convinced on this subject. I never understood that; I always thought that was a very weak point, honestly. My views now are not very different from what they were then. I was always very skeptical of that issue. Now, I would have to say you could imagine that the B cell receptor and the T cell receptor have a mimicry. It seems unlikely, but it could occur. Today I retain that skepticism. But I have been skeptical of a lot of things that I was wrong about. Skepticism can be a very bad idea (laughing), it keeps you from doing experiments.

Q6–Q7
Well certainly, this is the most interesting, of course. It is very difficult to look at the modern Treg field and the old suppressor T cells work, and to put them together. As you may know, I was a very close friend of Dick Gershon's. I remember the very first experiments that he did and the term he used, which is now in use again, "infectious tolerance". It was in the title of his first important paper[18]. I thought this was a very nice piece of work. But then he, and others, started developing these multiple-layered ideas in which idiotypy became very important in understanding specificity. I find that the modern suppressor work and the work done in the previous era are almost incompatible with each other. I think in a lot of those experiments there were so many potential variables that conclusions were being reached based on very modest support. The initial burst of work that indicated that there were powerful regulatory systems, I suspect was actually correct.

KE: Do you think that people got carried away with these ideas?
I absolutely think so. I still remember that there was an ingroup of people here, I don't know if it was true in Europe, but definitely here. Dick had started out as an outsider. Initially he was an outsider and his personality was most attractive as an outsider, trying to change the paradigm that people lived under. And then, he became an insider, an established figure, and that was bad for him. Dick wasn't "cut out" to be an establishment figure; it was bad for him. People were feeding data to him that they may have thought he wanted to see. I worked with Dick on a couple of things, I knew his technician, Kondo, who was really terrific. The work that Dick and Kondo did together was certainly right, I have great confidence in it.

KE: What was your attitude at the time?
My attitude was that the experiments were just too complex for us to do. There were so many variables that had to be controlled and so many reagents needed. However, I was impressed with many of the experiments. I couldn't see any holes in them but I did have major concerns.

Q8

We tried to study specific helper factors. As you know, I had been a postdoc with Baruj in New York City and we came to NIH I initially was quasi-independent. Half of my time I worked on my own things and I had first one and then two postdocs, and the rest of the time Baruj and I worked together and we had one postdoc in common. That was David Katz. In his first year, everything that David touched, worked. It was magnificent, every experiment David did worked. He published several very important papers. It was just at this time that people began to describe soluble suppressor and helper T cell factors that were specific. The factors could diffuse through filters. So in the second year David and I working together decided to attack this subject. Baruj had already moved to Harvard. We tried for a whole year to make progress on soluble specific helper factors. We never got one successful experiment. Experiment after experiment failed, while the year before everything David touched had worked. This sealed my admiration for David.

Q11

Definitely yes. I don't know how close you were to Niels. You know, in the US, Niels was not regarded in the same heroic light as he was in Europe. I have always felt that people tried to magnify his contributions to the clonal selection theory. I recently reviewed the subject and wrote a little piece for *Nature Immunology* on clonal selection. I admired Niels's ability to find issues of central importance, three in particular: First, he recognized that one could not explain the immune system based on instruction. Second, he grasped the central role of the thymus in generating a repertoire; and the third, he understood that the immune system required a potent regulatory mechanism. His contribution in the first case was to go back to Ehrlich and update his selection theory. I must say whether his reopening the selection idea helped Burnet or Talmage in moving to the view of clonal selection is hard for me to judge, but the Jerne theory of natural selection had very little to offer. Except for the concept of complexity of the repertoire, it went very little beyond Ehrlich. Second, his efforts on the thymus. He had a brilliant idea, but unfortunately it was incorrect. The notion that mutation occurred in the thymus was brilliant in the conception and needed to be considered. I regarded his ideas perhaps not as a revelation, but pretty darn important; unfortunately, his approach was wrong. And finally the network. I always believed that the network received much greater regard in Europe than in the US, partly because of that very high regard that Niels was held in Europe, as opposed to the more tempered enthusiasm for him here.

Q12

Eventually, and I must say, in my own mind, in the end the reason that I became less enthusiastic about the network, was that in the naive cell population, in contrast to the memory cells, the frequency of any individual

Hindsight

idiotype, except for the so-called public idiotypes, would be so low that an opportunity for cross-regulation would be limited by the very low likelihood that the cells regulating each other would succeed in finding each other, and that the concentrations of molecules in the blood would also be of too low a magnitude to be seriously involved in regulation. But I have to admit I really didn't think a lot about it. It was intellectually exciting, I became sufficiently interested so that Constantin Bona and I wrote this paper together, but I didn't work on the network experimentally and with time I felt that the data wasn't getting more and more compelling. The initial work was interesting but, in my own opinion there wasn't a continuing growth that made the data more and more compelling.

Q13
I still believe that surely there are instances in which there is recognition of idiotypes. The real question is not, does this occur? I don't think that's an important question, and the data are quite clear. The real question is, how central is it to the overall regulation and control of the immune system? And there my own position is, not very. But I never doubted that that idiotype-based regulation exists. If you had numbers, and you could put them into a way of modeling the system, you could ask: "Is this a central element of the determination of setting all the characteristics of the system?". My judgement today would be that its importance would be quite modest. There are instances, and probably they were found in appropriate experimental systems, for example when there are dominant idiotypes that were quite important in regulation. But the question would be how really major a component is that? Those instances which were used to form the basis of these experimental systems probably represent cases where they were really important. But I would think these were exceptions. I am judging, though, that many of my colleagues would be stronger in their opinion than I am.

Q14
The notion of T cell recognition of MHC in a sense intellectually superseded the network idea. It wasn't a competitive idea but, nonetheless, the notion of specificity in the T cell system diverted all the attention.

Q15
I don't know enough about it. I do know years ago there was the idea that you could use anti-idiotypes as vaccines. The concept there, as I understood it, was that there were cases in which it would be exceedingly difficult to prepare an antigenic vaccine, but in which you might be able to prepare the idiotype-mimic. It seems to me now that molecular biology has dealt with that problem, so in the vaccine area it's hard for me to see. Autoimmunity? I bet there would be examples, but I don't follow it carefully. In lupus, which I follow a little more, I know of the work of Edna Mozes with idiotypes.

KE: You know about the work of Dan Longo in non-Hodgkin lymphoma?
Yes, and of Ron Levy, but this is different. It is an entirely different idea, you can call it idiotypy, but you have a homogeneous population of cells, and the epitopes are reasonable candidates for making specific reagents against the tumor. Those epitopes are idiotypes, absolutely, but I see that as an entirely different approach, not related to the network. I see that as a good antigen to go after. Ron Levy had some success with non-Hodgkin's lymphoma, not a lot though. Anti-CD20 turned out to be far far more effective, and doesn't require individualized therapy. So the value as a drug that can be used in the clinic is so much greater.

Klaus Rajewsky, personal interview, Freiburg, Nov. 30, 2007.

Q1–Q4
My own connection to the network theory was like this: First, idiotypes were described by Oudin and Kunkel. Oudin's main point was that idiotypes were not inherited, they were individually specific. The reproducibility of this phenomenon was, so to speak, its non-reproducibility. And then there was the other side, people like you and Kunkel, who looked at idiotypes as genetic markers. I myself was fascinated by the Oudin papers. By the individual specificity. I became fascinated by immunology for reasons external to the immune system, because of things like memory, specificity, recognition, there was a philosophical/psychological interface. This was very important for me when I chose to work on it. I initially was not determined to go into science, but that strengthened my motivation enormously. So when Oudin came up with the specificity for the individual, I found it extremely fascinating, on that background. So I was primed for the network.

And then, there was of course Jerne. Niels became my scientific mentor. From him I learned what science is really all about. He had developed the natural selection theory (of antibody formation), in which everything starts with the antibodies which exist in the body before the antigen. Everything was due to these preexisting antibodies. He had missed the Burnetian idea of the one cell, one antibody rule. I know that Niels had a hard time to digest this, it had not occurred to him. Then when the idiotypes came up, antibodies were again in the center, he began to think of the immune system as an equilibrium between idiotypes and anti-idiotypes, and now all of a sudden the central problem of his original theory was solved in a new way, namely how come that when you inject an antigen you get more antibody of the kind that binds the antigen. In his original paper this is essentially mysterious and irrational, there was no real concept for this problem. But now he was again on top of Burnet in that again everything started from antibodies, which exist as an equilibrated network of idiotypes and anti-idiotypes, and the antibody response is just an adaptation to a new equilib-

Hindsight

rium. It was a theory proposed by a person who was, first of all, mainly a philosophical or mathematical thinker, and, secondly, there was this competition of ideas between him and mainly Burnet, of how the immune system might work. I was a victim of this because of my own preferences, as I just told you. I was extremely fascinated by the idiotypic network, as you know, and it took a while until I realized that it is somehow less central than I had thought.

Q5

In the case of the A5A idiotype, the genetic linkage of the T cell receptor idiotypes to the IgH allotype[19] was clear, wasn't it?

KE: Yes, the data were clear. As you remember, we tested this in experiments in which mice were injected with anti-idiotype and their spleen cells were used in microcultures as carrier-specific helper cells using StrepA as carrier. Induction of helper activity depended on the IgH allotype. We now know that T cell receptors are not encoded by VH, so the other possibility to explain the data is a selective influence in ontogeny of IgVH on the T cell repertoire. There are even very recent papers that this might be the case. I don't know if you know the work of Christina Joao[20]. She has a PCR-based method to look at the diversity of the T cell repertoire. In mice that have no B cells or just one clone of B cells the size of the T cell repertoire is ten-fold smaller. So this type of evidence seems to pop up again. She suggests that diverse immunoglobulins are required for generating a diverse T cell repertoire.

I still remember the dramatic effects, which we initially saw. When we did the first adoptive transfer experiments[21], I remember standing next to the counter seeing the gigantic anti-hapten responses. I was used to much lower responses in carrier-primed mice. The effects in anti-idiotype primed mice were quite incredible. I am convinced the original A5A experiments were real. We thought it was a real test to see what the nature of the receptor on T helper cells really was. Of course, in retrospect we know that we were fooled and that the evidence was indirect in several ways. I later wrote a paper on this, published in the *Scandinavian Journal of Immunology*, as an attempt to explain our results, to come to grips with our dilemma. There are two points: one, of course, is, as you just said, that immunoglobulin allotype linkage of T cell idiotypes does not necessarily imply that the latter are genetically identical to immunoglobulin idiotypes. The second point is, and that is something which for me was an important lesson, that the assays were all based on serology, and serology is just one of the most complex, difficult, and error-prone systems, it can fool you, even if you are extremely critical. When you have a DNA sequence, you can look at it any way you want, it is always either this or that. Serology is different. The results are always very much dependent on the way you do the test, which antibodies you use, does the second batch

164 Chapter 14

of antibodies give you the same result, and so on. It's a lot of things that make it unreliable.

KE: Can I ask you about your nylon fiber adsorption experiments of the T cell receptor?[22]
I think the main problem with these experiments was the tiny amounts of material. This is also something I had to learn. The tiny amounts of material were the basic problem, in combination with the extremely sensitive assay. And then, using serology to identify the nature of the minute amounts of material. It is simply so that whatever result we would have got, it would in the end have led to nowhere. It was just not something that you could put your hands on. Nowadays if you tell this to anybody in the scientific community they would say: 'It's a nice experiment, but get out enough material to do the molecular characterization'. It was 40 years ago, it was just impossible to do that kind of analysis.

Nowadays, in retrospect, I think that this approach to look at receptors was bound to fail because we had not enough material. It was a real mistake. I wouldn't dream of doing an experiment like this any more. Because it is not a way to find out what a molecule really is.

There were some people that had already started working with nucleic acids, but in practice this was not the kind of thing that people in immunology did. When Tonegawa came to Basel to study antibody diversity he told Niels that the only way to do it was to look at the DNA – Niels, who was of the opinion that by just thinking cleverly about things you could figure out how something would work. But in biology there is just not this logic. For Niels it was one of the biggest disappointments when Tonegawa found that immunoglobulin genes were organized in pieces, something that one would never have predicted.

I am still puzzling what these nylon purified receptors were. The phage inhibition assay was fine, the serology was as good as serology can be, the genetic linkage was fine as well. What we found with B cells were antibodies, you could precipitate them with anti-constant-region antibodies. In the case of T cells you could not do this, but they could be absorbed with anti-idiotype. What was it? Degraded antibodies? I don't know. If you want to use the adaptation argument, you could use it for the nylon-purified receptors as well.

Q6–Q8
I think these were all artifacts. Today there is solid evidence for suppressor cells, but these complicated circuits, all these factors! It was the same problem, no molecular methods, small amounts of material, small effects.

Q11
I think there were some lost souls in America, Heinz Köhler was one of those, I don't know exactly. I came to the US rather later, but most people there think it was a fundamental error, it was a wrong concept.

Q12

It is interesting to think about how come that at some point I just stopped to work on the idiotypic network. Many of us did. For me this is quite simple. We used idiotypes and anti-idiotypes to study antibody repertoires, on the basis of idiotypic specificities. We had data that, when the immune response develops, you lose common idiotypes and get diversification. I thought this was indicating somatic mutation of antibodies. This was not what Niels liked, when we talked at Pasteur about this, he said Klaus, be careful. He thought this was against the principle of clonal selection. In the end we wrote several papers about these things, but in retrospect the interesting part of them is the molecular biology, not the idiotypic analysis. In fact, at one point we wrote a paper for *Cell*, where we said that once you have primed cells, in the secondary response you don't get somatic mutation any more, just expansion of the somatically mutated clones. Half of that paper was initially idiotypic analysis, but in the course of writing it we threw out entirely the idiotypic part, and in the end it became essentially a molecular paper, showing how somatic mutation accompanies class switching, and so on. So, I stopped working on idiotypes because the molecular biology became so much more telling and more interesting. The idiotypic experiments came to a dead end. We showed, all of us, that everything you can imagine to happen in the idiotypic network, you can make happen; but the physiological impact of these reactions remained elusive.

Let me ask you a question. I always thought that the idiotype suppressor cells that you had, which could be transferred, on and on, that was one of the most stunning experiments in this field[23]. These were T cells which clearly, as far as the data looked, suppressed a particular idiotype, just one idiotype, and didn't touch the rest. These were amazing cells. So how come that in your case, this work somehow died out?

KE: These were indeed amazing cells. I don't know about their nature, but Eli Sercarz told me recently about some Treg clones which seem to recognize idiotope peptides. I might have accidentally stumbled upon this type of T cell. But I have also very clear reasons why I stopped working on the A5A idiotype. One reason came from your lab, and that were the 7 monoclonal anti-StrepA antibodies that you made, none of which was reactive with the anti-idiotype. Then I was very much irritated when I learned that other groups had difficulties to reproduce the A5A system in their laboratories, for example Roger Perlmutter or Joe Davie. The other crossreactive idiotypic systems, such as T15 and ARS, these could be handled in many labs. The crossreactive A5A idiotype did not seem to be a robust experimental system, it depended perhaps on the guinea pig reagents that we made, which must have detected some broadly crossreactive idiotypic determinant. From my genetic experiments I was convinced that the A5A idiotype was a defined V gene product, so I didn't like these results at all.

166 Chapter 14

Q13

In the NP system the crossreactive idiotype was recognized by monoclonal anti-idiotype. Also, the monoclonal anti-idiotypes had immunoregulatory properties, you injected small amounts and you changed the responses. I think that was all solid work, but in the end it didn't lead anywhere, because nobody came up with a ... it was just not the real trigger for an immune response. From my perspective, it also petered away because it discredited what Niels liked to call a 'facile idea', namely the idea that the immune system was there to fight infection. The 'facile idea' was discarded, it was negated, the system was all working in itself. This just turned out to be the wrong concept, not very productive.

When you think about the A5A story, the strongest point was that your data showed that there was more than one antibody V gene. I remember at the end of one lecture you said just that. You had a crossover, and you said there must be at least two genes. That was absolutely right, and it was the first evidence. There was discussion about this after your talk, but at the end you said there must be at least two antibody V genes, and that was exactly right, very solid.

Q15

I think you are quite right here when you call this attempts. As we all showed in the old times, everything can work. If you get something to work, it's fine. You know there is this famous therapy (IVIg) of Kazatchkine in Paris. There was a lot of discussion on it, and indirect evidence that it had to do with immunoglobulin idiotypes. But now I think Ravetch has shown it is all due to the Fc part, it has nothing to do with the V region. It works even much better if you just use the Fc part.

KE: One has to wait. I came to the conclusion that scientific evidence turns into hard fact only with time, it needs to stand the trial of strength in other laboratories and in practical application. I analyzed the discovery of DNA as the principle of heredity by Avery, where one can see this very clearly.

Sometimes though things work in a different way, take the double helix as an example. There was this one paper by Watson and Crick which immediately suggested how the genetic material could be reduplicated and transmitted from one cell to the next. This became so overwhelmingly clear that it was immediately accepted. This sometimes also happens in situations where the experimental evidence is not really very good. For example, what I found very impressive is Burnet's book (The clonal selection theory of acquired immunity). It is just a theory, basically, but when it appeared there was not even much discussion, it was so apparent that this was how things had to work. In this sense Burnet's book had something in common with the initial paper on the double helix, although it was not based on direct experimental evidence. It explained in one stroke how antigen induces specific antibodies and how tolerance to self-antigens is induced. Before that

Hindsight

we had no idea. When this simple idea came about, everybody just knew that was it. Burnet himself was frightened. He was frightened to publish it, actually. Initially he published it in a remote place, some Australian journal. I read the book from the beginning to end. I found it extremely impressive. Burnet really looked at the literature, he quotes everything. Then he says there is some evidence which doesn't fit. But that is always so, there is always some contradictory evidence. But the evidence altogether very strongly suggests that this is how it works. It's a great book, actually. A very different world from that of Niels. His network hypothesis had a strong philosophical connection, was as close to Kierkegaard as to lymphocytes. Niels was also the total opposite to Av (Mitchison), my second mentor.

David Sachs, by telephone, July 30, 2007.

Q1–Q4
The staphylococcal nuclease crossreactive idiotypic system was my first interest in idiotypes but then I got into anti-MHC idiotypes as well. In my paper[24] I pretty much summarized my feelings about the whole MHC anti-idiotypes. And I haven't changed my position. I got in before Wigzell started that, I was already in it because I had done the staphylococcal nuclease before[25]. When the Wigzell stuff (producing anti-idiotypic antisera to MHC-reactive T cells in responder x stimulator F1 rats[26]) came out I tried to reproduce it but wasn't able to. I took the sera over to Wigzell's lab and tried it there but wasn't able to get it to work. This poor Japanese postdoc made all the sera, he made over 100 attempts, I took all the sera with me. At that point I wasn't able to figure out what was going on. When Wigzell first got in, they were going to clarify what was wrong with Ramseier's experiments, you know, the soluble antigen recognition structures determined in the PAR assay. Wigzell's stuff was supposed to be a lot clearer but unfortunately they only had one good serum that worked, and once they ran out of that serum I don't think anything that they were doing then worked either. Their good serum, 1003, was completely depleted when I got there. We had done everything in the mouse until then but since it did not work in the mouse we turned to Lewis and DA rats to reproduce Wigzell's experiments. I was so sure it was going to work and when it didn't I went to Sweden, I frequently go to Sweden anyway. Wigzell was extremely kind about this, he set up the lab for me, but it never worked. I do not know what was really going on in those studies.

What worked fine (in my hands) was anti-idiotype to anti-MHC antibodies, this worked fine. This was what Jerne's theory originally was about anyway. I don't think that anybody has ever disproved that anti-idiotype to antibodies have major effects. It's only at the level of the T cell that it has been a problem. I also don't think that anybody has ever figured out why things that worked then had not been reproducible but I don't think this questions the idiotype network theory in general, rather the specifics of certain experiments.

KE: Do you think that in these experiments mistakes have been made?
This is always the default, but it would be nice to figure out what they are. When something cannot be reproduced, either you are not doing it exactly right, or it was wrong. It all remains possible until someone disproves something, and I don't think anything has been disproved, only not confirmed.

Q5
Well I think there is no question that the TCR has similarities to Ig but it is not Ig. There is also a similarity between the genes involved but they are clearly different genes. So the experiments suggesting Ig receptors on T cells were wrong. This falls in the same area as we discussed before, but here it has been disproved.

Q6, Q7
Here I think that many of the experiments undoubtedly were correct, but the interpretation was wrong. The newly revised idea about regulatory T cells can probably explain the experiments that were right. Clearly the ideas of suppressors, contrasuppressors and suppressors of suppressors were based on a lot of hypotheses. It had to be hypothetical because there weren't any good markers and no well-characterized cells.

Q8
In regard to I-J again I can't believe that all the data were wrong. Certainly the interpretations were wrong because there is no I-J region. The interpretation probably depended upon that people were studying the (mixed) effects of more than one I region structure, for example I-A and I-E, and the simplest idea was to say that there was a new region I-J. The antisera, 3R anti-5R, they had to be wrong. Nobody ever found out what they reacted with. The reason for that is that what people assume are congenics are no longer really congenic, they have been maintained as inbred strains for some time, and there is drift between inbred strains in the background. So they originally started out as identical, but we know there is drift in inbred strains. So it is possible that by immunizing between congenic strains they could have raised antibodies to other factors. That may explain some of the data, but the experiments were probably not all wrong. So people built new theories on the basis of hypotheses that had never been proven.

Q9, Q10
It was stimulatory at first but became inhibitory thereafter, there was almost a dark age where people were afraid to mention suppression. Then it came back finally because there is such a thing as regulation. Suppressor cells became so much discredited that they were called the S-word[27]. In the

Hindsight 169

beginning it stimulated a lot of research. I definitely think in the long run the stimulatory influence was probably more predominant. Now I think it even is coming back, people realize that there really are regulatory influences between cell populations.

KE: You mentioned regulation, do you think that some of immune regulation is idiotypic?
There is no solid evidence but in my thinking it is still likely. The clonotype of a T cell could be a target on which some regulation occurs. It is like many things, if you jump too quickly. It's like when an exciting new clinical result comes out, everybody thinks ok we are now curing (say) cancer. Then when, of course, people keep dying of cancer, people say it was all wrong. It wasn't wrong, it's just that they overinterpreted it, they were overenthusiastic about the result. Too much hype, not enough solid data.

Q11
I think it was more accepted in Europe than in the US. Jerne had a major influence and I think much of that stimulated Cazenave and friends. It took off in Europe. There certainly was a lot in the US too, but more in Europe.

Q12, Q13
It disappeared because, as I said, it was overenthusiastically interpreted at the beginning. People wanted to extend it beyond the antibody, where it really has been demonstrated and nobody ever disproved it. They wanted to extend it to the T cell and, in doing so, they forced it so much and with so much enthusiasm, and then the results were irreproducible. Therefore it disappeared. This doesn't mean it was wrong.

Q14
It you are talking about antibody, I think the idiotypic network theory still exists. It isn't considered as important any more because we know so much about the influence of T cell immunity on antibody regulation, for example the T-dependent and T-independent antibody responses. This has become more important in the understanding of antibody responses than the network theory. But when you use the network properly, you can make good use of it. For instance we have shown that if we use anti-idiotypic antibodies to anti-MHC monoclonals, you can stimulate responses to that MHC. But the paradigms that have superseded it have been the T cell subsets including the regulatory T cells.

Q15
There are a lot of speculations on the role of idiotype (in clinical conditions) but I do not think anything has been proven. But I still think that there is a likely possibility that idiotype-antiidiotype or clonotype-anticlonotype (recognition) are involved and could be exploited for clinical approaches.

170   Chapter 14

KE: You work in transplantation, do you have projects going on with idiotypic background?
When I see the word idiotype in a paper I read it because I am interested. Whenever I have a result that is potentially interpretable on the basis of idiotype I try to look into it. But I can't say that I have any active research going on right now.

Eli Sercarz, written response, Jan. 3, 2008.

Q1
I was very favorably disposed toward network theory in the early and mid-70s. Partly, it was because of the theoretical beauty of Jerne's formulation. I was teaching a large class of undergraduates at UCLA and introduced the topic into the curriculum, using the early experiments to describe the possible usefulness of the network.

Q2
No, I didn't get involved directly in idiotypic research until about 1977.

Q3
We definitely were inspired by Jerne's work. However, we wanted to study T cell recognition of idiotypic determinants and the nature of public idiotypes.

Q4
No, we didn't reject INT then.

Q5
We had gathered some evidence that the public idiotype of the secondary antibody response to HEL could also be found on T suppressor cells[28]. We didn't think that this meant that the Ts receptor was Ig, but that there was a cross-reactive recognition of the idiotypic determinant(s) on the Ts receptor.

Q6
One thing that is generally overlooked in the discussion about the former research on suppressor T cells (Ts) is that it concerned CD8 T cells for the most part. There were too many papers showing the suppressive nature of such cells to believe it was a gigantic artifact. Our studies were repeatable by a few generations of trainees (a generation being about 3 years), and largely were directed towards CD8 Ts that were activated by particular epitopes either on native proteins, or in the form of peptides from these proteins[29]. In conclusion, I believe those older studies that demonstrate suppression of CD4 T cells and the B cells dependent on such help.

Hindsight

Q7
We performed many studies describing CD4 T cells that were necessary to collaborate with, or induce, CD8 Ts. They were called T suppressor inducers (Tsi) and we showed that these Tsi were different from "ordinary" Th. Most strikingly, the Tsi were I-J+ and could be killed with anti I-J alloantisera in a thoroughly convincing and repeatable fashion[29] – believe it or not! We never learned exactly what the Tsi induced although we believed it activated APC.

Q8
We never worked on factors and had a hard time believing some of the stories about their constitution and specificity. It would be instructive to repeat some of these studies with current methodologies.

Q9
I think its effect was quite positive – it surely wasn't an inhibitory force. Of course, to prove that something could happen in an (experimental) immune system was not proof that it really would ever take place in the unperturbed system – (e.g. Ab5 and Ab6, etc.).

Q10
Many investigations can be considered a waste of time and lead to no new insights into the workings of the system. That can be said of some idiotypic studies also.

Q11
Probably, because of the proximity to Jerne, Urbain, Rajewsky and yourself, I think that the European environment was more hospitable to idiotypic thinking. Americans tend to be more practical rather than theoretical and soon moved on to other topics such as the TCR, tolerance etc. Perhaps the finding that the TCR had no Ig components had a part to play in making idiotypic studies less popular (also see 12.)

Q12
Again, it's a matter of moving on to new topics, as one has seen in the last decade with its jump onto bandwagons such as innate immunity, Toll-like receptors, and the incredible Treg fixation.

Q13
I don't think that the INT ever was proven to be wrong or irrelevant. It is surely to some extent correct. We have been working on a system of feedback suppression involving CD8 Ts, requiring CD4 collaboration, in which the CD8 T cells are restricted by class Ib Qa-1 molecules which recognize a TCR receptor "idiopeptide" in its groove[30]. All the components have been identified and it represents a system type which has not yet been exploited

sufficiently. Despite parts of this system having been described in the literature for more than 10 years, it is hard to get a fair hearing: this is a holdover from the old CD8 days of disbelief. However, I have confidence that one day, this will be recognized to be one common methodology of regulation. It is also a direct consequence of idiotypic thinking.

Q14

I think that innate immunity and autoimmunity are very broad areas which excited much interest, along with examining lymphocyte dynamics with fluorescent dyes and direct visualization, the Hollywoodization of immunology. Detection of T cell specificity via tetramers was another technical victory enjoyed by many. But the idiotypic network theory was such an exciting and beautiful concept, a harking back to the fantastic era of Monod, Jacob and the Pasteur group, that it brings with it a whole aura of pleasure at contemplating and arguing about a whole Weltanschauung.

Q15

I'd like to have another lifetime to devote to such studies. I'd also love to have your answers to these 15 questions.

Jacques Urbain, by telephone, July 24, 2007.

Q1–Q4

My position is a little bit special because I came to network ideas independently of the Jerne papers. I started immunology in 68/69, at the end of instructive theories. I was thinking about selective theories and there was this strong opposition between somatic and germline theories. I do not remember why but I did not like somatic mutation. I was preferring germline. Of course for germline theories the problem was to imagine a selective pressure which was sufficient to keep antibodies, for example against crocodile albumin, in the mouse or the rabbit. I imagined if the diversity is so great there must be idiotype self-recognition in the system, and this could provide the selective pressure to keep a large number of genes.

Furthermore, my first work in immunology was studying the evolution of binding affinity in the immune response. Using an antigen which does not need any adjuvant, tobacco mosaic virus, I found that the initial increase in binding affinity was followed by a decrease. In thinking about the reasons I was in need of a mechanism that distinguished between high and low affinity antibodies. It did not seem to be antigen because at the end of the response the antigen concentration should be low so that high-affinity antibodies should predominate. So I imagined that there was some kind of idiotype recognition. I published this only in 5 lines in a paper on cellular recognition in evolution[31]. My thinking at that time was very much inspired by a

Hindsight

paper of Niels, not the network papers but a pre-network paper of 1971: "What precedes clonal selection?"

When I heard the famous talk of Jerne at Pasteur in 1974 on idiotypic network I took this as a boost because I had felt alone in my thinking on idiotypes. I realized that the ideas were extremely similar to my own. I then became really a fan of the network hypothesis. Believing in the germline theory I thought that all idiotypes are present in all individuals of a species, so that one should be able to turn private idiotypes into public ones. That was the beginning of the Ab1, Ab2, Ab3... idiotypic cascade[32].

Q5

With respect to the Ig nature of the T cell receptor I like to distinguish between serology and the functional data. As far as serology is concerned, I think that some of the antisera against Ig recognized glycan side chains. I remember working with an anti-KLH antiserum that perfectly stained T cells due to anti-glycan antibodies. I was not the only one having these experiences so my interpretation was that these were anti-sugar antibodies. The second possibility which to my knowledge has never been discarded is that T cell receptors and immunoglobulins are after all evolutionarily related structures derived from a common ancestor. It is in fact not impossible that some antibodies recognize common structures. I do not know any study to establish that but it is not impossible. Of course there may have been some other artifacts. For functional data you are better placed than me to comment. I have been very much impressed by recent data by Maria Cascalho and Christina Joao[20,33], in which it is shown that in mice devoid of B cells, for example J-H KO mice or the monoclonal mice of Mathias Wabl, the repertoire of T cells is very much reduced, at least by a factor of 10. If all of this is correct, the interpretation is that the T cell repertoire depends on the repertoire of B cells. Positive selection of T cells could be influenced by idio-peptides of B cells. This would also explain the VH linkage of T cell idiotypes.

Q6

With respect to suppressor T cells, I always thought that they are existing. I do not know of any mechanism in biology without a negative feedback. I understand the experimental difficulties but I do not understand why people completely rejected the notion of suppressor T cells, on the basis of I-J and so on. In fact, suppressor T cells are enjoying a vivid revival.

Q7

The Th1/Th2 concept, that was the killing of the network. If I look back 10–15 years ago, everything was interpreted in terms of Th1/Th2. The dogma was Th1/Th2 and no suppressor T cells. Now we have Th1/Th2, Th17, and Treg which are of course suppressor T cells. I had always some doubts about this. In the Mossmann paper about 20 years ago they studied clones but if you look in the normal immune response most of the T cells are not polarized, there are a lot of T cells which look like what they called Th0.

KE: I did not mean the Th1/Th2 concept here, I meant the inducer/effector concept for helper and suppressor T cells.
The concept of inducer and effector T cell for both help and suppression I discussed a lot with Marc Greene. In fact, he told me that 2/3 of the experiments did not work, it is not in the papers but this is what he told me.

Q8
With antigen-specific helper and suppressor factors I always had some difficulties. This must have been complex mixtures of cytokines and so on. I do not understand how they got reproducible results with such complex mixtures. I have not much to say about factors.

Q9, Q10
The network theory was very stimulating, indeed is still stimulating today. Ok, the wrong results were inhibitory. It would be stupid to say that it was only inhibitory. Some was useful and some was a waste.

Q11
I think there was a difference between Europe and the US. Europe was much more in favor than the US. In the US it was only a small group, like the lab of Bill Paul, Constantin Bona, Don Capra. But several conferences had been held in the US. New ideas are always more easily welcomed in Europe. Also the funding system in the US works against novel developments.

Q12, Q14
This is a very difficult but interesting question. I can see three reasons. The first is the sharing of idiotypes between T and B cells and the discovery of the non-immunoglobulin nature of T cell receptors. This did not kill but put the network into suspicion. The second is more surprising. You know about the Imanishi/Baltimore case[34], which is now even the subject of a 500-page book. For several months the *New York Times* published articles at least once per week about the case on the front page. Normally you would not expect the *New York Times* to have an interest in idiotype research. I discussed with many people that this case had a strong negative effect on idiotype research, especially in the States. The presence of the transgenic idiotype induces the appearance of endogenous related idiotypes. There is a technical problem with the specificity of one reagent. However, there are other works from other labs which establish clearly idiotype mimicry and imprinting either by maternal idiotypes or by injection of monoclonal antibodies. To my knowledge these facts are extremely difficult to understand without idiotype networks.

Hindsight   175

The third reason was, as I said above, the Th1/Th2 concept. This was the paradigm that followed the network. This I never understood because of some older work, of Parish and others, showing that there were antagonistic effects between delayed type hypersensitivity (DTH) and antibody synthesis. If you have good antibody synthesis, you get no DTH, if you have a good DTH, no antibody synthesis. This was never properly explained by the Th1/Th2 concept.

The network even disappeared from the textbooks. The 5th edition of the Janeway textbook did not mention the network. The 6th edition mentions it again because Janeway himself had published a paper in support of the network in the *PNAS* just before he died.

Q13
As far as regulation is concerned, perhaps in the network there was some confusion between recognition and action. Network is concerned with recognition, not action. Action is due to cytokines, most probably. I am convinced that to some extent the theory is correct in terms of repertoire selection.

In parentheses, let me talk for one minute about one of my papers, the best paper I think I ever wrote. It was published in the *European Journal of Immunology* 1995[35]. The idea was to test the network hypothesis by removing the auto-Ab2 B lymphocytes to see if there was an influence on the idiotypic repertoire in mice. We used Nisonoff's ARS major idiotypic system in AJ mice. We prepared syngeneic Ab2 in mice and prepared Ab3, i.e. anti-Ab2, in rabbits, because rabbit antibodies are good binders of complement. Then we injected the Ab3 into mice to get rid of auto-Ab2. Then we immunized the mice with the antigen. The result was that the anti-ARS response was good, but the antibodies did not posses the crossreactive idiotype. I think nobody ever read it, it was never cited, it was not very amusing. The network was already dead.

Q15
Clinical research is the last pocket of resistance, there are still many papers. I am convinced that there could be some interesting developments. Recently I was reading about certain rare antibodies that can be protective against HIV. Perhaps by using idiotype vaccination it is possible to induce such special antibodies in the human. If I had money and people I would get involved in this project.

Hans Wigzell, by telephone, July 26, 2007.
The network theory is to some extent an example for the influence of sociology in science. I gave a lecture in the Montreal Immunology Congress, and I introduced also an element of personality. An interesting thing is the link of suppressor T cells to Dick Gershon. Suppressor cells then died and

176 Chapter 14

then came regulatory T cells. They died because they were not a physical entity. There was a difference between what materialized into well-defined biophysical entities or very well-defined cells versus suppressor T cells which was more of a phenomenological or descriptive term. The association of idiotypes with Niels Jerne is not so clear cut as that of suppressor cells with Dick Gershon.

Q1, Q2
My attitude in the beginning, it was positive. But I was involved in idiotypic research before that. I read a lot of Jacques Oudins articles and I went to visit him in Paris. Then I came in contact with Jean Lindenmann. I was already involved, not directly in idiotypic network theory but kind of in it, in a sense. It made a lot of sense to me from the point of view, let's say, that these specificities could be recognized like antigens by a variety of bio-chemical/biophysical parameters reflected by the antibodies. I was not amazed by these ideas, I thought this was logical. But some people took it to the extreme, like Niels Jerne, as a completely dominating thing. That Niels believed in the dominance of the idiotypic network is documented by the fact that he refused blood transfusion when he was operated of his prostate cancer. You know why? He thought that it would ruin his network. He almost died. The guy was a missionary, a firm believer. Isn't this interesting?

Q3, Q4
I was interested because of the signaling of the mother to the child. I think that it is a too simplistic view to think that the child is born with a virgin immune system. This is clearly not the case. We did some work on this, it has been modulated.

I did not reject the network, it is one of the ways in which one can regulate the immune system, we know that. I have been involved in giving monoclonal antibodies to two people with specificity to cancer antigens and antiidiotypic responses turn up, with an interesting spread between T and B cell epitopes. So there is something in it, ok?

Q5
If you look on it, it was a fascinating thing because there was two absolutely diametrically contradictory things. One was this very rigid notion of MHC-linked suppressor factors which were supposed to be coded for by some genes in the MHC. At this time there was no evidence that there was this extreme heterogeneity in the individual, right? Then there were a lot of people finding Ig-like molecules, Marchalonis and so on. And we, Hans Binz and I, we were trying to find out the genetics underlying it[26]. I am still actually not really sure what the hell we did find[36].

KE: Do you agree that the VH allotype linkage most probably was a mistake?
We were working mostly in the rat, in the GVH system and the MHC. And

Hindsight

we also found a linkage (to heavy chain allotype) and we could verify that by breeding. We made the (anti-idiotypic) sera in F1 hybrids, we had a low frequency of animals that were able to make these sera, but if we had a serum it was beautiful. The readout was not a conventional antibody response but the blocking of GVH. I have not looked into it but what might have played a role was that the NK-linked receptors are extremely important for the hybrid histocompatibility reaction, and the blockage of the GVH of that type. I do not know if what we found were any of these KIR-linked molecules or whatever.

KE: What do you think of the possibility that the T cell repertoire is educated by the B cell repertoire?
I think it's true, definitely true. I think it was solid research from the point of view of methodology but I think there was something more with respect to the linkage between the T cell specificity and the B cell specificity, absolutely.

The molecule that we got out was not a normal immunoglobulin molecule nor did it match any of the conventional T cell receptors (known today). Out of rat urine we could extract a molecule which was of about 70 K molecular weight[36]. It was an extremely exciting and an extremely frustrating time. I still think there are some missing elements in that regard. The controversy then was emotional and the people coming in, Mark Davis and Tak Mak, who stumbled on it, what they found was comparatively trivial from the point of chemistry. But the recognition of MHC/peptide was of course conceptually quite distinct from antibodies.

Q6
As I was alluding to in the beginning, this suffered very much from the fact that there was no way of isolating these cells, there was a lack of good markers. What Dick Gershon and his group were doing was excellent experimental research but the interpretations of the data that he got, suppressors and counter-suppressors and the like, were too simplistic, sort of. I found a fantastic quote that I used the other day. When Miller and Mitchell first presented their data on T and B cells, they were heavily attacked "for introducing unnecessary complications in the immune system". Isn't that a good one? It shows that in those times these were very emotional issues. When Eva Klein, Rolf Kissling and I found these natural killer cells, we suffered a lot. Although NK cells account for several percent of the cells in the blood, we were considered to have a hole in the head. The only support we got was from some German hematologist, I do not remember his name, who had seen these cells and talked about some granular lymphocytes. This was based on morphology, and I remember how attacked he was for several years.

Q7
This is in the same ballpark as Q6. These cells were actually poorly defined.

There was also very little knowledge about the cytokines or chemokines, and markers. It was some kind of descriptive science trying to simplify things maybe a bit too much.

Q8

With respect to the soluble helper and suppressor factors I am still lost. Tomio Tada's IgE suppressor factor and the like, I think we were left in the dark.

Q9

I think it had a very stimulatory effect. If you look at immunology today, much of what is being considered successful has actually created practical results, clinical results. The most obvious ones are technology based, such as the monoclonal antibodies. With respect to vaccination the idiotypic approach leads to a very restricted kind of response. The Ab2, Ab3 kind of thing, you could actually repeat this even in man, there is no doubt about it. So I think it may have been overheated at one time, but right now it is significantly underheated. You are considered a radical if you stand up and say that immunoglobulin going through the placenta leaves an imprint in the immune system of the child. The new generation wouldn't even know what you are talking about. I think this is a great mistake.

Q10

I think this was at a breakpoint where you had still room for theories. If you like, now there is too little theory. It is molecular and boring. It is fact collection and it is very poorly put into an integrative biological function. It is not that one is getting old, it is really boring. The immunological articles are more or less like CNN reports, very fragmented, short incomplete stories. There is time now for a deeper analysis, right?

You and the other Klaus (Rajewsky) had this system of streptococcal immunization of mice, and it showed how each mouse made different antibodies even in the same litter, like the butterfly chaos kind of thing. It was not a waste of time, it was overinterpreted maybe, particularly by Niels Jerne as a sort of dominating concept.

KE: Would you say that the regulatory role of the network was overestimated?
Yes, but I would say that it is the fine tuning, the fact that the fetus is born, lets say, with a kind of semi-primed immune profile. I do not know who did this experiment, he immunized mothers with the O antigen of *E. coli* and showed that the offspring resisted an *E. coli* infection. I think if you could inject mothers with an appropriate monoclonal antibody you could do something very nice for these neonatally dangerous infections.

Q11

Europe was certainly dominating. It is an interesting question, I think we perhaps have space for more theory. But you could turn this around and say that the suppressor T cells came from the US. They were certainly wild in many ways, in a conceptual way. But from the point of view of impact (in network research), there was more from Europe. But I may be biased.

Q12, Q13, Q15
It disappeared, in part, because it was difficult to study, it remains fragmentary, except in select quarters like the impact on the fetus. Also it did not deliver, for example from the point of view of vaccines. People like Gus Dalgleish still are trying in the field of cancer vaccines, we have been involved with melanoma antigens and CEA structures, but it hasn't delivered with a clinical endpoint. This is part of it. Take the EU situation there is no doubt that it has to be good science, but then the value of its practical possibilities, the weight of that has been increasing profoundly in the granting situation. This kind of more esoteric stuff would actually then suffer. This is why it suffered, not for other reasons.

Q14
The antigen-presenting cells in their complexity, plus or minus signals, the danger concept, Polly (Matzinger)/ Charley Janeway kind of quarrels, these kind of things. Many of these developments were actually technology driven.

180   Chapter 14

References and further reading, Part II

Chapter 6

References

1   Krause RM (1981) *The restless tide; the persistent challenge of the microbial world*. The National Foundation for Infectious Diseases, Washington
2   Silverstein AM (1989) *A history of immunology*. Academic Press, Inc. San Diego, New York
3   Fleck L (1979) *Genesis and Development of a Scientific Fact*, Ed. by. T.J. Trenn and R.K. Merton. The University of Chicago Press, Chicago/London
4   Kindt T, Capra D (1984) *The antibody enigma*. Plenum Press, New York, London
5   Von Behring EA, Kitasato S (1890) Über das Zustandekommen der Diphterieimmunität bei Thieren. *Dtsch Med Wochenschrift* 16: 1113
6   Ehrlich P (1900) On immunity with special reference to cell life. *Proc R Soc London (Biol)* 66: 424
7   Bäumler E (1984) *Paul Ehrlich, scientist for life*. Holmes and Meier Publ., New York, London
8   Ehrlich P (1897) Zur Kenntnis der Antitoxinwirkung. *Fortschr D Med* 15: 41
9   Tiselius A, Kabat EA (1939) An electrophoretic study of immune sera and purified antibody preparations. *J Exp Med* 69: 119–131
10 Landsteiner K (1922) On the formation of heterogenetic antigen by combination of hapten and protein. *Proc of the Koninklijke Akademie van Weetenschappen te Amsterdam* 24 (1/7): 237
11 Pauling L (1940) A theory of the structure and process of the formation of antibody. *J Am Chem Soc* 62: 2643
12 Breinl F, Haurowitz F, (1930) Chemische Untersuchung das Präzipitates aus Hämoglobin und Anti-hämoglobin-serum und Bemerkungen über die Natur der Antikörper. *Z Phys Chem* 192: 45
13 Avery OT, MacLeod C, McCarty M, (1944) Studies on the chemical nature of the substance inducing transformation of pneumococcal types. *J Exp Med* 79: 137–158
14 Watson JD, Crick FH (1953) Molecular structure of nucleic acids; a structure for deoxyribose nucleic acid. *Nature* 171: 737–738
15 Edelman GM, Gally JA, (1964) A model for the 7S antibody molecule. *Proc Nat Acad Sci USA* 51: 846
16 Porter RR (1967) The structure of antibodies. *Sci American* 217: 81–87
17 Jerne NK (1955) The natural selection theory of antibody formation. *Proc Nat Acad Sci USA* 41: 849
18 Söderquist T (2003) *Science as autobiography, the troubled life of Niels Jerne*. Yale University Press, New Haven, London
19 Jerne N (1951) *A study of avidity based on rabbit skin responses to diphteria toxin-antitoxin mixtures*. Munksgaard, Copenhagen
20 Jerne N (1956) The presence in normal serum of specific antibody against bacteriophage T4 and its increase during the earliest stages of immunization. *J Immunol* 76: 209
21 Burnet FM (1957) A modification of Jerne's theory of antibody production using the concept of clonal selection. *Aust J Sci* 20: 67

References and further reading, Part II

22   Burnet FM (1959) *The clonal selection theory of aquired immunity*. Cambridge University Press, Cambridge
23   Lederberg J (1959) Genes and antibodies. Do antigens bear instructions for antibody specificity or do they select cell lines that arise by mutation? *Science* 129: 1649–1653
24   Talmage DW (1959) Immunological specificity, unique combinations of selected natural globulins provide an alternative to the classical concept. *Science* 129: 1643–1648
25   Chase M (1945) The cellular transfer of cutaneous hypersensitivity to tuberculin. *Proc Soc Exp Biol* 59: 134
26   Mitchison NA (1953) Passive transfer of transplantation immunity. *Nature* 171: 267
27   Gowans JL, McGregor DD, Cowen DM (1962) Initiation of Immune Responses by Small Lymphocytes. *Nature* 196: 651
28   Fragreus A (1948) *Acta Med Scand* Suppl 204
29   Coons AH, Leduc EH, Connally JM (1955) Studies on antibody production. I. A method for the histochemical demonstration of specific antibody *J Exp Med* 102: 49
30   Nossal GJV (1958) Antibody production by single cells. *Brit J Exp Physiol* 39: 544
31   Ehrlich P, Morgenroth J (1901) Über Hämolysine. V. Mitteilung. *Berl Klin Wochenschr* 10: 251
32   Donath J, Landsteiner K, (1904) Ueber paroxysmale Hämoglobinurie. *Münch Med Wochenschr* 51: 1590
33   Burnet FM, Fenner F (1949) *The production of antibodies*. 2nd ed. MacMillan, New York
34   Billingham RE, Brent L, Medawar PB (1953) Actively acquired tolerance of foreign cells. *Nature* 172: 603–606
35   Weigle WO (1961) The immune response of rabbits tolerant to bovine serum albumin to the injection of other heterologous serum albumins. *J Exp Med* 114: 111
36   Wilson WE, Talmage DW (1965) Erythrocyte chimerism and acquired immunological tolerance. *J Immunol* 94: 150–156
37   Kunkel HG (1965) Myeloma proteins and antibodies. *Harvey Lect* 59: 219
38   Natvig JG, Kunkel HG (1973) Human immunoglobulins: Classes, subclasses, genetic variants, and idiotypes. *Adv Immunol* 16: 1
39   Hilschmann N, Craig LC (1965) Amino acid sequence studies with Bence-Jones proteins. *Proc Nat Acad Sci USA* 53: 1403

Further reading

Claman HN (1963) Tolerance to a protein antigen in adult mice and the effect of non-specific factors. *J Immunol* 91: 833

Ohno S (1970) *Evolution by gene duplication*. Springer Verlag, New York

Landsteiner K (1947) *The specificity of serological reactions*. Harvard University Press, Cambridge, Massachusetts

Jerne HK, Nordin AA (1963) Plaque formation in agar by single antibody-producing cells. *Science* 140: 405

Potter M (1972) Immunoglobulin-producing tumors and myeloma proteins in mice. *Physiol Rev* 52: 631

Chapter 7

References

1   Chase M (1945) The cellular transfer of cutaneous hypersensitivity to tuberculin. *Proc Soc Exp Biol* 59: 134

2   Mitchison NA (1953) Passive transfer of transplantation immunity. *Nature* 171: 267

3   Miller JFA (2003) Biological curiosities, what can we learn from them? In: Eichmann K (ed): *The biology of complex organisms. Creation and protection of integrity*. Birkhäuser, Basel

4   Miller JFA (1961) Immunological function of the thymus. *Lancet* 2: 748–749

5   Claman HN, Chaperon EA, Triplett RF (1966) Immunocompetence of transferred thymus-marrow cell combinations. *J Immunol* 97: 828–832

6   Claman HN, Chaperon EA, Triplett RF (1966) Thymus-marrow cell combinations. Synergism in antibody production. *Proc Soc Exp Biol Med* 122: 1167–1171

7   Miller JFA, Mitchell GF (1967) Thymus and precursors of antigen reactive cells. *Nature* 216: 659

8   Miller JFA, Mitchell GF (1968) Cell to cell interaction in immune response.1. Hemolysis-forming cells in neonatally thymectomized mice reconstituted with thymus or thoracic duct lymphocytes. *J Exp Med* 128: 801

9   Mitchell GF, Miller JFA (1968) Cell to cell interaction in immune response. 2. Source of hemolysin-forming cells in irradiated mice given bone marrow and thymus or thoracic duct lymphocytes. *J Exp Med* 128: 821

10   Raff MC (1969) Theta isoantigen as a marker of thymus-derived lymphocytes in mice. *Nature* 224: 378–379

11   Moller G (1961) Demonstration of mouse isoantigens at cellular level by fluorescent antibody technique. *J Exp Med* 114: 415

12   Raff MC (1970) Two distinct populations of peripheral lymphocytes in mice distinguishable by immunofluorescence. *Immunology* 19: 637–650

13   Raff MC, Sternberg M, Taylor RB (1970) Immunoglobulin determinants on the surface of mouse lymphoid cells. *Nature* 225: 553–554

14   Ford WL, Gowans J (1969) The traffic of lymphocytes. *Semin Hematol* 6: 67–83

15   Cantor H, Boyse EA (1975) Development and function of subclasses of T cells. *J Reticuloendothel Soc* 17: 115–118

16   Weissman IL, Masuda T, Olive C, Friedberg SH (1975) Differentiation of migration of T lymphocytes. *Isr J Med Sci* 11: 1267–1277

17   Schlesinger M (1970) Anti-theta antibodies for detecting thymus-dependent lymphocytes in the immune response of mice to SRBC. *Nature* 226: 1254–1256

18   Playfair JH, Purves EC (1971) Antibody formation by bone marrow cells in irradiated mice. I. Thymus-dependent and thymus-independent responses to sheep erythrocytes. *Immunology* 21: 113–121

19   Rajewsky K, Schirrmacher V, Nase S, Jerne NK (1969) The requirement of more than one antigenic determinant for immunogenicity. *J Exp Med* 129: 1131–1143

References and further reading, Part II 183

20 Mitchison NA (1971) The carrier effect in the secondary response to hapten-protein conjugates. II. Cellular cooperation. *Eur J Immunol* 1: 18–27

21 Ilya Ilyich Metchnikov: *Intra-Cellular Digestion* (1882), *The Comparative Pathology of Inflammation* (1892), and *Immunity in Infectious Diseases* (1905)

22 Zarafonetis C, Harmon DR, Clark PF (1947) The influence of temperature upon opsonization and phagocytosis. *J Bacteriol* 53: 343–349

23 Mishell RI, Dutton RW (1967) Immunization of dissociated spleen cell cultures from normal mice. *J Exp Med* 126: 423–442

24 Askonas BA, Auzins I, Unanue ER (1968) Role of macrophages in the immune response. *Bull Soc Chim Biol (Paris)* 50: 1113–1128

25 Unanue ER, Cerottini JC (1970) The function of macrophages in the immune response. *Semin Hematol* 7: 225–248

26 Bach FH, Voynow NK (1966) One-way stimulation in mixed leukocyte cultures. *Science* 153: 545–547

27 Wilson DB (1967) Quantitative studies on the mixed lymphocyte interaction in rats. I. Conditions and parameters of response. *J Exp Med* 126: 625–54

28 Simonsen M (1965) Recent experiments on the graft-*versus*-host reaction in the chick embryo. *Br Med Bull* 21: 129–32

29 Nisbet NW, Simonsen M, Zaleski M (1969) The frequency of antigen-sensitive cells in tissue transplantation. A commentary on clonal selection. *J Exp Med* 129: 459–467

30 Klein J (1966) Strength of some H-2 antigens in mice. *Folia Biol (Praha)* 12: 168–175

31 Bach FH (1968) Transplantation: problems of histocompatibility testing. *Science* 159: 1196–1198

32 Bach FH, Amos DB (1967) Hu-1: Major histocompatibility locus in man. *Science* 156: 1506–1508

33 Davies DA, Manstone AJ, Viza DC, Colombani J, Dausset J (1968) Human transplantation antigens: the HL-A (Hu-1) system and its homology with the mouse H-2 system. *Transplantation* 6: 571–586

34 Ceppelini R (1971) Old and new facts and speculations about transplantation antigens of man. Progr Immunol I, Amos B. Ed. Academic Press, p. 973

35 Hayry P, Defendi V (1970) Mixed lymphocyte cultures produce effector cells: model *in vitro* for allograft rejection. *Science* 168: 133–135

36 Hayry P, Andersson LC (1973) T cells in mixed-lymphocyte-culture-induced cytolysis (MLC-CML). *Transplant Proc* 5: 1697–703

37 Cerottini JC, Engers HD, Macdonald HR, Brunner T (1974) Generation of cytotoxic T lymphocytes *in vitro*. I. Response of normal and immune mouse spleen cells in mixed leukocyte cultures. *J Exp Med* 140: 703–717

38 Andersson LC, Hayry P (1973) Specific priming of mouse thymus-dependent lymphocytes to allogeneic cells *in vitro*. *Eur J Immunol* 3: 595–599

39 Andersson LC, Nordling S, Hayry P (1973) Allograft immunity *in vitro*. VI. Autonomy of T lymphocytes in target cell destruction. *Scand J Immunol* 2: 107–113

40 Gardner I, Bowern NA, Blanden RV (1974) Cell-mediated cytotoxicity against ectromelia virus-infected target cells. I. Specificity and kinetics. *Eur J Immunol* 4: 63–67

41 McDevitt HO, Sela M (1965) Genetic control of the antibody response. I. Demonstration of determinant-specific differences in response to synthetic polypeptide antigens in two strains of inbred mice. *J Exp Med* 122: 517–531

184 Chapter 14

42 McDevitt HO, Sela M (1967) Genetic control of the antibody response. II. Further analysis of the specificity of determinant-specific control, and genetic analysis of the response to (H,G)-A–L in CBA and C57 mice. *J Exp Med* 126: 969–978

43 Ellman L, Green I, Martin WJ, Benacerraf B (1970) Linkage between the poly-L-lysine gene and the locus controlling the major histocompatibility antigens in strain 2 guinea pigs. *Proc Natl Acad Sci USA* 66: 322–328

44 Zinkernagel RM, Doherty PC (1974) Restriction of *in vitro* T cell-mediated cytotoxicity in lymphocytic choriomeningitis within a syngeneic or semiallogeneic system. *Nature* 248: 701–702

45 Zinkernagel RM (1974) Restriction by H-2 gene complex of transfer of cell-mediated immunity to *Listeria monocytogenes. Nature* 251: 230–233

46 Kindred B (1975) The failure of allogeneic cells to maintain an immune response in nude mice. *Scand J Immunol* 4: 653–656

47 Katz DH, Graves M, Dorf ME, Dimuzio H, Benacerraf B (1975) Cell interactions between histoincompatible T and B lymphocytes. VII. Cooperative responses between lymphocytes are controlled by genes in the I region of the H-2 complex. *J Exp Med* 141: 263–268

48 Claman HN, McDonald W (1964) Thymus and X-irradiation in the termination of aquired immunological tolerance in the adult mouse. *Nature* 202: 713
Claman HN, Talmage DW (1963) Thymectomy: Prolongation of the immunological tolerance in the adult mouse. *Science* 141: 1193

49 Droege W (1971) Amplifying and suppressive effect of thymus cells. *Nature* 234: 549–551

50 Gershon RK, Kondo K (1971) Infectious immunological tolerance. *Immunology* 21: 903–914
Gershon RK, Cohen P, Hencin R, Liebhaber SA. 1972. Suppressor T cells. *J Immunol* 108: 586–590

51 Gershon RK (1975) A disquisition on suppressor T cells. *Transplant Rev* 26: 170–185

52 Green DR, Flood PM, Gershon RK (1983) Immunoregulatory T cell pathways. *Ann Rev Immunol* 1: 439

53 Jerne NK, (1971) The somatic generation of immune recognition. *Eur J Immunol* 1: 1

54 Bretscher P, Cohn M (1970) A theory of self-nonself discrimination. *Science* 169: 1042

55 Langman RE, Cohn M (1993) Two signal models of lymphocyte activation? *Immunol Today* 14: 235–237

Further reading

Good RA, Finstad J (1968) The development and involution of the lymphoid system and immunologic capacity. *Trans Am Clin Climatol Assoc* 79: 69–107
Cohn M (1970) Selection under a somatic model. *Cell Immunol* 1: 461–467
Cohn M (1972) Immunology: what are the rules of the game? *Cell Immunol* 5: 1–20
Van Rood JJ, Van Leeuwen A, Freudenberg J, Rubinstein P (1971) Prospects in host-donor matching. *Transplant Proc* 3: 1042–50
Herzenberg LA, Chan EL, Ravitch MM, Riblet RJ, Herzenberg LA. (1973) Active suppression of immunoglobulin allotype synthesis. 3. Identification of T cells as

References and further reading, Part II 185

responsible for suppression by cells from spleen, thymus, lymph node, and bone marrow. *J Exp Med* 137: 1311–1324

Gershon RK, Mokyr MB, Mitchell MS (1974).Activation of suppressor T cells by tumour cells and specific antibody. *Nature* 250: 594–596

Kirkwood JM, Gershon RK (1974) A role for suppressor T cells in immunological enhancement of tumor growth. *Prog Exp Tumor Res* 19: 157–164

Chapter 8

References

1. Fleck L (1979) *Genesis and Development of a Scientific Fact*, ed. by Trenn TJ, Merton RK. The University of Chicago Press, Chicago/London
2. Krause RM (1981) Would the Wassermann reaction survive peer review? *Conference on Bacterial Virulence and Pathogenesis*. Rocky Mountain Laboratory, NIAID, Mamilton MO. Unpublished manuscript
3. Uhr JW, Baumann JB (1961) Antibody formation. I. The suppression of antibody formation by passively administered antibody. *J Exp Med* 113: 935–957
4. Franklin EC, Kunkel HG (1957) Immunologic differences between the 19 S and 7 S components of normal human gamma-globulin. *J Immunol* 78: 11–18
5. Finkelstein MS, Uhr JW (1964) Specific inhibition of antibody formation by passively administered 19S and 7S antibody. *Science* 146: 67–69
6. Henry C, Jerne NK (1968) Competition of 19S and 7S antigen receptors in the regulation of the primary immune response. *J Exp Med* 128: 133–152
7. Moller G, Wigzell H (1965) Antibody sysnthesis at the cellular level: Antibody induced suppression of 19S and 7S antibody response. *J Exp Med* 121: 969–989
8. Jerne NK (1955) The natural selection theory of antibody formation. *Proc Nat Acad Sci USA* 41: 849
9. Burnet FM (1957) A modification of Jerne's theory of antibody production using the concept of clonal selection. *Aust J Sci* 20: 67
10. Burnet FM (1959) *The clonal selection theory of aquired immunity*. Cambridge University Press, Cambridge
11. Kunkel HG (1968) The "abnormality" of myeloma proteins. *Cancer Res* 28: 1351–1353
12. Kunkel HG (1965) Myeloma proteins and antibodies. *Harvey Lect* 59: 219
13. Slater RJ, Ward SM, Kunkel HG (1955) Immunological relationships among the myeloma proteins. *J Exp Med* 101: 85–108
14. Grey HM, Mannik M, Kunkel HG (1965) Individual antigenic specificity of myeloma proteins. Characteristics and localization to subunits. *J Exp Med* 121: 561–575
15. Oudin J (1960) Allotypy of rabbit serum proteins. II. Relationships between various allotypes: their common antigenic specificity, their distribution in a sample population; genetic implications. *J Exp Med* 112: 125–142
16. Grubb R (1965) Human gamma-globulin polymorphism. *Bibl Haemato* 23: 375–382
17. Herzenberg LA, McDevitt H, Herzenberg LA (1968) Genetics of antibodies. *Ann Rev Genet* 2: 209
18. Todd CW (1972) Genetic control of H chain biosynthesis in the rabbit. *Fed Proc* 31: 188–192

19 Oudin J, Michel M (1969) Idiotypy of rabbit antibodies. I. Comparison of idiotypy of antibodies against *Salmonella typhi* with that of antibodies against other bacteria in the same rabbits, or of antibodies against *Salmonella typhi* in various rabbits. *J Exp Med* 130: 595–617

20 Oudin J, Michel M (1969) Idiotypy of rabbit antibodies. II. Comparison of idiotypy of various kinds of antibodies formed in the same rabbits against *Salmonella typhi*. *J Exp Med* 130: 619–642

21 Kunkel HG, Mannik M, Williams RC (1963) The individual antigenic specificity of isolated antibodies. *Science* 140: 1218

22 Eichmann K (1975) Genetic control of antibody specificity in the mouse. *Immunogenetics* 2: 491–506

23 Capra JD, Kehoe JM (1975) Hypervariable regions, idiotypy, and the antibody combining site. *Adv Immunol* 20: 951

24 Oudin J, Cazenave PA (1971) Similar idiotypic specificities in immunoglobulin fractions with different antibody functions or even without detectable antibody function. *Proc Natl Acad Sci USA* 68: 2616–2620

25 Brient BW, Nisonoff A (1970) Quantitative investigations of idiotypic antibodies, IV. Inhibition by specific haptens of the reaction of antihapten antibody with its anti-idiotypic antibody. *J Exp Med* 132: 951

26 Nisonoff A, Lamoyi E. (1981) Implications of the presence of an internal image of the antigen in anti-idiotypic antibodies: possible application to vaccine production. *Clin Immunol Immunopathol* 21: 397–406

27 Jerne NK (1974) Towards a network theory of the immune system. *Ann Immunol (Inst Pasteur)* 125 C. 373

28 Uhr JW, Baumann JB (1961) Antibody formation. II. The specific anamnestic antibody response. *J Exp Med* 113: 959–70

29 Singer W (1999) Hirnforschung an der Schwelle zum nächsten Jahrhundert. In: Sitte P (ed): *Jahrhunderstwissenschaft Biologie.* C.H.Beck'sche Verlagsbuchhandlung, München

30 Wiener N (1948/1961) *Cybernetics, or control of communication in the animal and the machine.* MIT Press, Cambridge, MA

31 Rosenbluth A, Wiener N, Bigelow J (1943) Behavior, purpose, and teleology. *Philosophy of Science* 10: 18

32 Wiener N (1956) *The human use of human beings: Cybernetics and society.* Doubleday, Garden City, New York

33 Von Foerster H (1974) On constructing a reality. In: Von Foerster (ed): *Observing systems.* Intersystem Publications, Salinas, CA

34 Van Bertalanffy L (1975) *General systems theory, Foundations, Developments, Applications.* Braziller, New York

35 Maturana H, Varela F (1980) *Autopoesis and cognition: The realization of the Living.* Reidel, Dordrecht, Boston

36 Geyer F, Van der Zwouten (eds) (1990) *Self-referencing in social systems.* Intersystem Publications, Salinas, CA

Further reading

Eichmann K, Kindt TJ (1971) The inheritance of individual antigenic specificities of rabbit antibodies to streptococcal carbohydrates. *J Exp Med* 134: 532–552

Eichmann K, Tung AS, Nisonoff A (1974) Linkage and rearrangement of genes encoding mouse immunoglobulin heavy chains. *Nature* 250: 509–511

Penn GM, Kunkel HG, Grey HM (1970) Sharing of individual antigenic determinants between a gamma G and a gamma M protein in the same myeloma serum. *Proc Soc Exp Biol Med* 135: 660–665

Natvig JG, Kunkel HG (1973) Human immunoglobulins: Classes, subclasses, genetic variants, and idiotypes. *Adv Immunol* 16: 1

Oudin J (1956) The allotype of certain blood protein antigens. *C R Hebd Seances Acad Sci* 242: 2606–2608 (in French)

Oudin J (1960) Allotypes of certain serum protein antigens. Immuno-chemical and genetic relationships between 6 principal allotypes observed in rabbit serum. *C R Hebd Seances Acad Sci* 250: 770–772 (in French)

Oudin J, Michel M (1969) On the idiotypic specificity of rabbit anti-*S. typhi* antibodies. *C R Acad Sci Hebd Seances Acad Sci D* 268: 230–233 (in French)

Eichmann K (1978) Expression and function of idiotypes of lymphocytes. *Adv Immunol* 26: 195–254

Chapter 9

References

1   Jerne NK (1960) Immunological speculations. *Ann Rev Microbiol* 14: 341
2   Kunkel HG, Mannik M, Williams RC (1963) The individual antigenic specificity of isolated antibodies. *Science* 140: 1218
3   Slater RJ, Ward SM, Kunkel HG (1955) Immunological relationships among the myeloma proteins. *J Exp Med* 101: 85-108
4   Oudin J, Michel M (1969) Idiotypy of rabbit antibodies. I. Comparison of idiotypy of antibodies against *Salmonella typhi* with that of antibodies against other bacteria in the same rabbits, or of antibodies against *Salmonella typhi* in various rabbits. *J Exp Med* 130: 595–617
5   Sirisinha S, Eisen HN (1971) Autoimmune-like antibodies to the ligand-binding sites of myeloma proteins. *Proc Nat Acad Sci USA* 68: 3130
6   Janeway CA Jr, Paul WE (1973) Hapten-specific augmentation of the anti-idiotype antibody response to hapten-myeloma protein conjugates in mice. *Eur J Immunol* 3: 340
7   Rodkey LS (1974) Studies of idiotypic antibodies. Production and characterization of autoantiidiotypic antisera. *J Exp Med* 139: 712
8   Jerne NK (1974) Towards a network theory of the immune system. *Ann Immunol (Inst Pasteur)* 125 C: 373
9   Harrison MR, Mage RG (1973) Allotype suppression in the rabbit. I. The ontogeny of cells bearing immunoglobulin of paternal allotype and the fate of these cells after treatment with antiallotype antisera. *J Exp Med* 138: 764–774
10 Cosenza H, Kohler H (1972) Specific suppression of the antibody response by antibodies to receptors. *Proc Nat Acad Sci USA* 69: 1701
11 Hart DA, Wang AL, Nisonoff A (1972) Suppression of idiotypic specificities in adult mice by administration of antiidiotypic antibody. *J Exp Med* 135: 1293
12 Rowley DA, Fitch FW, Stuart FP, Köhler H, Cosenza H (1973) Specific suppression of immune responses. *Science* 181: 1133

13 Gershon RK, Cohen P, Hencin R, Liebhaber SA (1972) Suppressor T cells. *J Immunol* 108: 586–590

14 Herzenberg LA, Chan EL, Ravitch MM, Riblet RJ, Herzenberg LA (1973) Active suppression of immunoglobulin allotype synthesis. 3. Identification of T cells as responsible for suppression by cells from spleen, thymus, lymph node, and bone marrow. *J Exp Med* 137: 1311–1324

15 Jerne NK (1976) The immune system: a web of V-domains. *Harvey Lectures* 70: 93

16 Oudin J, Cazenave PA (1971) Similar idiotypic specificities in immunoglobulin fractions with different antibody functions or even without detectable antibody function. *Proc Nat Acad Sci USA* 68: 2616

17 Cohn M (1985) Diversity in the immune system: "preconceived ideas" or ideas preconceived? *Biochimie* 67: 9–27
Cohn M (1986) Is the immune system a functional idiotypic network? *Ann Inst Pasteur Immunol* 137C: 173–188
Cohn M (1986) The concept of functional idiotype network for immune regulation mocks all and comforts none. *Ann Inst Pasteur Immunol* 137C: 64–76
Cohn M (1981) Conversations with Niels Kaj Jerne on immune regulation: associative versus network recognition. *Cell Immunol* 61: 425–436

18 Kluskens L, Köhler H (1974) Regulation of immune response by autogenous antibody against receptor. *Proc Nat Acad Sci* 71: 5083

19 Cosenza H (1976) Detection of anti-idiotype reactive cells in the response to phosphorylcholine. *Eur J Immunol* 6: 114–116

20 McKearn TJ, Stuart FP, Fitch FW (1974) Anti-idiotypic antibody in rat transplantation immunity. I. Production of anti-idiotypic antibody in animals repeatedly immunized with alloantigens. *J Immunol* 113: 1876–1882

21 Eichmann K (1975) Genetic control of antibody specificity in the mouse. *Immunogenetics* 2: 491

22 Weigert M, Potter M (1977) Antibody varable region genetics: Summary and abstract of the homogeneous antibody workshop VII. *Immunogenestics* 4: 401

23 Greene MI, Nisonoff A (eds) (1984) *The biology of idiotypes*. Plenum Press, New York

24 Eichmann K (1974) Idiotype suppression. I. Influence of the dose and of the effector function of anti-idiotypic antibody on the production of an idiotype. *Eur J Immunol* 4: 296

25 Strayer DS, Cosenza H, Lee WMF, Rowley DA, Köhler H (1974) Neonatal tolerance induced by antibody against antigen-specific receptor. *Science* 186: 640

26 Augustin A, Cosenza H (1976) Expression of new idiotypes following neonatal idiotypic suppression of a dominant clone. *Eur J Immunol* 6: 497–501

27 Trenkner E, Riblet R (1975) Induction of antiphosphorylcholine antibody formation by anti-idiotypic antibodies. *J Exp Med* 142: 1121

28 Nisonoff A, Lamoyi E (1981) Implications of the presence of an internal image of the antigen in anti-idiotypic antibodies: possible application to vaccine production. *Clin Immunol Immunopathol* 21: 397–406

29 Takemori T, Tesch H, Reth M, Rajewsky K (1982) The immune response against anti-idiotope antibodies. I. Induction of idiotope-bearing antibodies and analysis of the idiotope repertoire. *Eur J Immunol* 12: 1040–1046

30 Gardner MR, Ashby WR (1970) Connectance of large dynamuic (cybernetic) systems: critical values for stability. *Nature* 228: 784

References and further reading, Part II 189

31 May RM (1972) Will a large complex system be stable? *Nature* 238: 413
32 Richter PH (1975) A network theory of the immune system. *Eur J Immunol* 5: 350
33 Hoffmann GW (1975) A theory of regulation and self-nonself discrimination in an immune network. *Eur J Immunol* 5: 638
34 Taussig MJ (1974) T cell factor which can replace T cells *in vivo. Nature* 248: 234
35 Adam G, Weiler E (1976) Lymphocyte population dynamics during ontogenic generation of diversity. In: Cunningham AJ (ed): *The Generation of Antibody Diversity*. Academic Press, New York, 1–20

Further reading

Jerne NK (1985) The generative grammar of the immune system. *EMBO J* 4: 847–852
Jerne NK (1984) Idiotypic networks and other preconceived ideas. *Immunol Rev* 79: 5–24
Jerne NK, Roland J, Cazenave PA (1982) Recurrent idiotopes and internal images. *EMBO J* 1: 243–247
Jerne NK (1977) The common sense of immunology. *Cold Spring Harb Symp Quant Biol* 41: 1
Jerne NK (1974) Clonal selection in a lymphocyte network. *Soc Gen Physiol Ser* 29: 39–48
Rajewsky K, Takemori T (1983) Genetics, expression, and function of idiotypes. *Annu Rev Immunol* 1: 569–607

Chapter 10

References

1 Jerne NK (1974–1975) The immune system: a web of V-domains. *Harvey Lect* 70, Series 93–110
2 Jerne NK (1974) Towards a network theory of the immune system. *Ann Immunol Inst Pasteur* 125C: 373
3 Benacerraf B, McDevitt HO (1972) Histocompatibility-linked immune response genes. *Science* 175: 273–279
4 Benacerraf B, Germain RN (1978) The immune response genes of the major histocompatibility complex. *Immunol Rev* 38: 70–119
5 Germain RN, Benacerraf B (1980) Helper and suppressor T cell factors. *Springer Semin Immunopathol* 93–127
6 Jerne NK (1971) The somatic generation of immune recognition. *Eur J Immunol* 1: 1
7 Raff MC (1970) Two distinct populations of peripheral lymphocytes in mice distinguishable by immunofluorescence. *Immunology* 19: 637–650
8 Roelants GE, Forni L, Pernis B (1973) Blocking and redistribution (capping) of antigen receptors on T and B lymphocytes by anti-immunoglobulin antibody. *J Exp Med* 137: 1060
9 McKearn TJ (1974) Antireceptor antiserum causes specific inhibition of reactivity to rat histocomparibility antigens. *Science* 183: 94

10   Marchalonis JJ, Decker JM, DeLuca D, Moseley J M, Smith P, Warr G W (1977) Lymphocyte surface immunoglobulins, evolutionary origin and involvement in activation. *Cold Spring Harb Symp Quant Biol* 41: 261
Du Pasquier L, Weiss N, Loor F (1972) Direct evidence for immunoglobulins on the surface of thymus lymphocytes of amphibian larvae. *Eur J Immunol* 2: 366–370
Ruben LN, Van der Hoven A, Dutton RW (1973) Cellular cooperation in hapten-carrier responses in the newt, *Triturus viridescens. Cell Immunol* 6: 300–314
Emmrich F, Richter RF, Ambrosius H (1975) Immunoglobulin determinants on the surface of lymphoid cells of carps. *Eur J Immunol* 5: 76–78
Yocum D, Cuchens M, Clem LW (1975) The hapten-carrier effect in teleost fish. *J Immunol* 114: 925–927

11   Eichmann K, Rajewsky K (1975) Induction of T and B cell immunity by anti-idiotypic antibody. *Eur J Immunol* 5: 661–666

12   Krawinkel U, Cramer M, Berek C, Hämmerling G, Black SJ, Rajewsky K, Eichmann K (1977) On the structure of the T cell receptor for antigen. *Cold Spring Harb Symp Quant Biol* 41: 285

13   Eichmann K (1978) Expression and function of idiotypes on lymphocytes. *Adv Immunol* 26: 195

14   Krawinkel U, Rajewsky K (1976) Specific enrichment of antigen-binding receptors from sensitized murine lymphocytes. *Eur J Immunol* 6: 529–536

15   Krawinkel U, Cramer M, Mage RG, Kelus AS, Rajewsky K (1977) Isolated hapten-binding receptors of sensitized lymphocytes. II. Receptors from nylon wool-enriched rabbit T lymphocytes lack serological determinants of immunoglobulin constant domains but carry the A locus allotypic markers. *J Exp Med* 146: 792–801

16   Krawinkel U, Cramer M, Imanishi-Kari T, Jack RS, Rajewsky K, Makela O (1977) Isolated hapten-binding receptors of sensitized lymphocytes. I. Receptors from nylon wool-enriched mouse T lymphocytes lack serological markers of immunoglobulin constant domains but express heavy chain variable portions. *Eur J Immunol* 7: 566–573

17   Binz H, Wigzell H, Bazin H (1976) T-cell idiotypes are linked to immunoglobulin heavy chain genes. *Nature* 264: 639–642

18   Binz H, Wigzell H (1977) Antigen-binding, idiotypic receptors from T lymphocytes: an analysis of their biochemistry, genetics, and use as immunogens to produce specific immune tolerance. *Cold Spring Harb Symp Quant Biol* 41: 275–284

19   Edelman GM (1977) Summary: understanding selective molecular recognition. *Cold Spring Harb Symp Quant Biol* 41: 891–902

20   Hedrick SM, Cohen DI, Nielsen EA, Davis MM (1984) Isolation of cDNA clones encoding T cell-specific membrane-associated proteins. *Nature* 308: 149–153

21   Yanagi Y, Yoshikai Y, Leggett K, Clark SP, Aleksander I, Mak TW (1984) A human T cell-specific cDNA clone encodes a protein having extensive homology to immunoglobulin chains. *Nature* 308: 145–149

22   Adam G, Weiler E (1976) Lymphocyte population dynamics during ontogenic generation of diversity. In: Cunningham AJ (ed): *The Generation of Antibody Diversity.* Academic Press, New York, 1–20

23   Rajewsky K (1984) Mechanisms of idiotypic control of the antibody repertoire.

References and further reading, Part II

In: Greene MI, Nisonoff A (eds): *The Biology of Idiotypes*. Plenum Press, New York, 477

24 João C (2007) Immunoglobulin is a highly diverse self-molecule that improves cellular diversity and function during immune reconstitution. *Med Hypotheses* 68: 158–161

25 Zinkernagel RM, Doherty PC (1974) Restriction of *in vitro* T cell-mediated cytotoxicity in lymphocytic choriomeningitis within a syngeneic or semiallogeneic system. *Nature* 248: 701–702

26 Zinkernagel RM (1974) Restriction by H-2 gene complex of transfer of cell-mediated immunity to *Listeria monocytogenes*. *Nature* 251: 230–233

27 Kindred B (1975) The failure of allogeneic cells to maintain an immune response in nude mice. *Scand J Immunol* 4: 653–656

28 Katz DH, Graves M, Dorf ME, Dimuzio H, Benacerraf B (1975) Cell interactions between histoincompatible T and B lymphocytes. VII. Cooperative responses between lymphocytes are controlled by genes in the I region of the H-2 complex. *J Exp Med* 141: 263–268

29 Krammer PH, Eichmann K (1977) T cell receptor idiotypes are controlled by genes in the heavy chain linkage group and the major histocompatibility complex. *Nature* 270: 733–735

30 Zinkernagel RM, Althage A, Cooper S, Callahan G, Klein J (1978) In irradiation chimeras, K or D regions of the chimeric host, not of the donor lymphocytes, determine immune responsiveness of antiviral cytotoxic T cells. *J Exp Med* 148: 805–810

31 Fremont DH, Matsumura M, Stura EA, Peterson PA, Wilson IA (1992) Crystal structures of two viral peptides in complex with murine MHC class I H-2Kb. *Science* 257: 919–927

32 Rothbard JB (1990) The ternary complex: T cell receptor, MHC protein, and immunogenic peptide. *Semin Immunol* 5: 283–295

33 Köhler G, Milstein C (1975) Continuous cultures of fused cells secreting antibody of predefined specificity. *Nature* 256: 495–497

34 Haskins K, Kubo R, White J, Pigeon M, Kappler J, Marrack P (1983) The major histocompatibility complex-restricted antigen receptor on T cells. I. Isolation with a monoclonal antibody. *J Exp Med* 157: 1149–1169

35 Staerz UD, Pasternack MS, Klein JR, Benedetto JD, Bevan MJ (1984) Monoclonal antibodies specific for a murine cytotoxic T-lymphocyte clone. *Proc Natl Acad Sci USA* 81: 1799–1803

36 Meuer SC, Fitzgerald KA, Hussey RE, Hodgdon JC, Schlossman SF, Reinherz EL (1983) Clonotypic structures involved in antigen-specific human T cell function. Relationship to the T3 molecular complex. *J Exp Med* 157: 705–719

37 Kronenberg M, Kraig E, Siu G, Kapp JA, Kappler J, Marrack P, Pierce CW, Hood L (1983) Three T cell hybridomas do not contain detectable heavy chain variable gene transcripts. *J Exp Med* 158: 210–227

Further reading

Doherty PC, Zinkernagel RM (1975) H-2 compatibility is required for T-cell-mediated lysis of target cells infected with lymphocytic choriomeningitis virus. *J Exp Med* 141: 502–507

Zinkernagel RM, Doherty PC (1977) The concept that surveillance of self is mediated *via* the same set of genes that determines recognition of allogenic cells. *Cold Spring Harb Symp Quant Biol* 41: 505–510

Roehm N, Herron L, Cambier J, DiGuisto D, Haskins K, Kappler J, Marrack P (1984) The major histocompatibility complex-restricted antigen receptor on T cells: distribution on thymus and peripheral T cells. *Cell* 38: 577–584

Marrack P, Hannum C, Harris M, Haskins K, Kubo R, Pigeon M, Shimonkevitz R, White J, Kappler J (1983) Antigen-specific, major histocompatibility complex-restricted T cell receptors. *Immunol Rev* 76: 131–145

Kronenberg M, Kraig E, Hood L (1983) Finding the T-cell antigen receptor: past attempts and future promise. *Cell* 34: 327–329

Hedrick SM, Nielsen EA, Kavaler J, Cohen DI, Davis MM (1984) Sequence relationships between putative T-cell receptor polypeptides and immunoglobulins. *Nature* 308: 153–158

McDevitt HO (2000) Discovering the role of the major histocompatibility complex in the immune response. *Annu Rev Immunol* 18: 1–17

Benacerraf B (1978) A hypothesis to relate the specificity of T lymphocytes and the activity of I region-specific Ir genes in macrophages and B lymphocytes. *J Immunol* 120: 1809–1812

Chapter 11

References

1   Droege W (1971) Amplifying and suppressive effect of thymus cells. *Nature* 234: 549–551

2   Gershon RK, Kondo K (1971) Infectious immunological tolerance. *Immunology* 21: 903–914

3   Gershon RK, Cohen P, Hencin R, Liebhaber SA (1972) Suppressor T cells. *J Immunol* 108: 586–590

4   Herzenberg LA, Chan EL, Ravitch MM, Riblet RJ, Herzenberg LA (1973) Active suppression of immunoglobulin allotype synthesis. 3. Identification of T cells as responsible for suppression by cells from spleen, thymus, lymph node, and bone marrow. *J Exp Med* 137: 1311–1324

5   Jerne NK (1974) Towards a network theory of the immune system. *Ann Immunol (Inst Pasteur)* 125C: 373

6   Raff MC (1970) Two distinct populations of peripheral lymphocytes in mice distinguishable by immunofluorescence. *Immunology* 19: 637–650

7   Cantor H, Boyse EA (1975) Functional subclasses of T lymphocytes bearing different Ly antigens. II. Cooperation between subclasses of Ly+ cells in the generation of killer activity. *J Exp Med* 141: 1390–1399

8   Cantor H, Boyse EA (1975) Functional subclasses of T-lymphocytes bearing different Ly antigens. I. The generation of functionally distinct T-cell subclasses is a differentiative process independent of antigen. *J Exp Med* 141: 1376–1389

9   Ledbetter JA, Herzenberg LA (1979) Xenogeneic monoclonal antibodies to mouse lymphoid differentiation antigens. *Immunol Rev* 47: 63–90

10 Ledbetter JA, Rouse RV, Micklem HS, Herzenberg LA (1980) T cell subsets

defined by expression of Lyt-1,2,3 and Thy-1 antigens. Two-parameter immunofluorescence and cytotoxicity analysis with monoclonal antibodies modifies current views. *J Exp Med* 152: 280–295

11   Huber B, Devinsky O, Gershon RK, Cantor H (1976) Cell-mediated immunity: delayed-type hypersensitivity and cytotoxic responses are mediated by different T-cell subclasses. *J Exp Med* 143: 1534–1539

12   Cantor H, Simpson E (1975) Regulation of the immune response by subclasses of T lymphocytes. I. Interactions between pre-killer T cells and regulatory T cells obtained from peripheral lymphoid tissues of mice. *Eur J Immunol* 5: 330–336

13   Cantor H, Boyse EA (1975) Development and function of subclasses of T cells. *J Reticuloendothel Soc* 17: 115–118

14   Jandinski J, Cantor H, Tadakuma T, Peavy DL, Pierce CW (1976) Separation of helper T cells from suppressor T cells expressing different Ly components. I. Polyclonal activation: suppressor and helper activities are inherent properties of distinct T-cell subclasses. *J Exp Med* 143: 1382–1390

15   Cantor H, Gershon RK (1979) Immunological circuits: cellular composition. *Fed Proc* 38: 2058–2064

16   Green DR, Flood PM, Gershon RK (1983) Immunoregulatory T-cell pathways. *Annu Rev Immunol* 1: 439–463

17   Tada T, Taniguchi M, David CS (1976) Properties of the antigen-specific suppressive T-cell factor in the regulation of antibody response of the mouse. IV. Special subregion assignment of the gene(s) that codes for the suppressive T-cell factor in the H-2 histocompatibility complex. *J Exp Med* 144: 713–725

18   Murphy DB, Herzenberg LA, Okumura K, Herzenberg LA, McDevitt HO (1976) A new I subregion (I-J) marked by a locus (Ia-4) controlling surface determinants on suppressor T lymphocytes. *J Exp Med* 144: 699–712

19   Tada T, Okumura K (1979) The role of antigen-specific T cell factors in the immune response. *Adv Immunol* 28: 1–87

20   Munro AJ, Taussig MJ, Campbell R, Williams H, Lawson Y (1974) Antigen-specific T-cell factor in cell cooperation: physical properties and mapping in the left-hand (K) half of H-2. *J Exp Med* 140: 1579–1587

21   Taussig MJ (1974) T cell factor which can replace T cells *in vivo. Nature* 248: 234–236

22   Tada T, Okumura K, Taniguchi M (1973) Regulation of homocytotropic antibody formation in the rat. 8. An antigen-specific T cell factor that regulates anti-hapten homocytotropic antibody response. *J Immunol* 111: 952–961;
Takemori T, Tada T (1975) Properties of antigen-specific suppressive T-cell factor in the regulation of antibody response of the mouse. I. *In vivo* activity and immunochemical characterization. *J Exp Med* 142: 1241–1253;
Treves AJ, Cohen IR, Feldman M (1976) Suppressor factor secreted by T-lymphocytes from tumor-bearing mice. *J Natl Cancer Inst* 57: 409–414;
Thomas DW, Roberts WK, Talmage DW (1975) Regulation of the immune response: production of a soluble suppressor by immune spleen cells *in vitro. J Immunol* 114: 1616–1622;
Rich SS, Rich RR (1976) Regulatory mechanisms in cell-mediated immune responses. IV. Expression of a receptor for mixed lymphocyte reaction suppressor factor on activated T lymphocytes. *J Exp Med* 144: 1214–1226;
Harwell L, Marrack P, Kappler JW (1977) Suppressor T-cell inactivation of a helper T-cell factor. *Nature* 265: 57–59;

Theze J, Kapp JA, Benacerraf B (1977) Immunosuppressive factor(s) extracted from lymphoid cells of nonresponder mice primed with L-glutamic acid60-L-alanine30-L-tyrosine10 (GAT) III Immunochemical properties of the GAT-specific suppressive factor. *J Exp Med* 145: 839–856;

Germain RN, Theze J, Waltenbaugh C, Dorf ME, Benacerraf B (1978) Antigen-specific T cell-mediated suppression. II. *In vitro* induction by I-J-coded L-glutamic acid50-L-tyrosine50 (GT)-specific T cell suppressor factor (GT-T8F) of suppressor T cells (T82) bearing distinct I-J determinants. *J Immunol* 121: 602–607;

Kontiainen S, Simpson E, Bohrer E, Beverley PC, Herzenberg LA, Fitzpatrick WC, Vogt P, Torano A, McKenzie IF, Feldmann M (1978) T-cell lines producing antigen-specific suppressor factor. *Nature* 274: 477–480;

Perry LL, Benacerraf B, Greene MI (1978) Regulation of the immune response to tumor antigen. IV. Tumor antigen-specific suppressor factor(s) bear I-J determinants and induce suppressor T cells *in vivo*. *J Immunol* 121: 2144–2147

23 Germain RN, Benacerraf B (1980) Helper and suppressor T cell factors. *Springer Semin Immunopathol* 3: 93–127

24 Eichmann K, Rajewsky K (1975) Induction of T and B cell immunity by anti-idiotypic antibody. *Eur J Immunol* 5: 661–666

25 Sy MS, Bach BA, Dohi Y, Nisonoff A, Benacerraf B, Greene MI (1979) Antigen- and receptor-driven regulatory mechanisms. I. Induction of suppressor T cells with anti-idiotypic antibodies. *J Exp Med* 150: 1216–1228

26 Sy MS, Bach BA, Brown A, Nisonoff A, Benacerraf B, Greene MI (1979) Antigen- and receptor-driven regulatory mechanisms. II. Induction of suppressor T cells with idiotype-coupled syngeneic spleen cells. *J Exp Med* 150: 1229–1240

27 Eichmann K (1975) Idiotype suppression. II. Amplification of a suppressor T cell with anti-idiotypic activity. *Eur J Immunol* 5: 511–517

28 Woodland R, Cantor H (1978) Idiotype-specific T helper cells are required to induce idiotype-positive B memory cells to secrete antibody. *Eur J Immunol* 8: 600–606

29 Hetzelberger D, Eichmann K (1978) Recognition of idiotypes in lymphocyte interactions. I. Idiotypic selectivity in the cooperation between T and B lymphocytes. *Eur J Immunol* 8: 846–852

30 Eichmann K, Falk I, Rajewsky K (1978) Recognition of idiotypes in lymphocyte interactions. II. Antigen-independent cooperation between T and B lymphocytes that possess similar and complementary idiotypes. *Eur J Immunol* 8: 853–857

31 Dietz MH, Sy MS, Benacerraf B, Nisonoff A, Greene MI, Germain RN (1981) Antigen- and receptor-driven regulatory mechanisms. VII. H-2-restricted anti-idiotypic suppressor factor from efferent suppressor T cells. *J Exp Med* 153: 450–463

32 Sy MS, Nisonoff A, Germain RN, Benacerraf B, Greene MI (1981) Antigen- and receptor-driven regulatory mechanisms. VIII. Suppression of idiotype-negative, p-azobenzenearsonate-specific T cells results from the interaction of an anti-idiotypic second-order T suppressor cell with a cross-reactive-idiotype-positive, p-azobenzenearsonate-primed T cell target. *J Exp Med* 153: 1415–1425

33 Takaoki M, Sy MS, Whitaker B, Nepom J, Finberg R, Germain RN, Nisonoff A,

Benacerraf B, Greene MI (1982) Biologic activity of an idiotype-bearing suppressor T cell factor produced by a long-term T cell hybridoma. *J Immunol* 128: 49–53

34 Greene MI, Pierres A, Dorf ME, Benacerraf B (1977) The I-J subregion codes for determinats on suppressor factor(s) which limit the contact sensitivity response to picryl chloride. *J Exp Med* 146: 293–296

Greene MI, Benacerraf B (1980) Studies on hapten specific T cell immunity and suppression. *Immunol Rev* 50: 163–186

Minami M, Okuda K, Furusawa S, Benacerraf B, Dorf ME (1981) Analysis of T cell hybridomas. I. Characterization of H-2 and Igh-restricted monoclonal suppressor factors. *J Exp Med* 154: 1390–1402

Takaoki M, Sy MS, Tominaga A, Lowy A, Tsurufuji M, Finberg R, Benacerraf B, Greene MI (1982) I-J-restricted interactions in the generation of azobenzenearsonate-specific suppressor T cells. *J Exp Med* 156: 1325–1334

Okuda K, Minami M, Ju ST, Dorf ME (1981) Functional association of idiotypic and I-J determinants on the antigen receptor of suppressor T cells. *Proc Natl Acad Sci USA* 78: 4557–4561

Sherr DH, Dorf ME (1984) Characterization of anti-idiotypic suppressor T cells (Tsid) induced after antigen priming. *J Immunol* 133: 1142–1150

35 Flood PM, Lowy A, Tominaga A, Chue B, Greene MI, Gershon RK (1983) Igh variable region-restricted T cell interactions. Genetic restriction of an antigenspecific suppressor inducer factor is imparted by an I-J+ antigen-nonspecific molecule. *J Exp Med* 158: 1938–1947

36 Eardley DD, Shen FW, Cantor H, Gershon RK (1979) Genetic control of immunoregulatory circuits. Genes linked to the Ig locus govern communication between regulatory T-cell sets. *J Exp Med* 150: 44–50

37 Flood PM, DeLeo AB, Old LJ, Gershon RK (1983) Relation of cell surface antigens on methylcholanthrene-induced fibrosarcomas to immunoglobulin heavy chain complex variable region-linked T cell interaction molecules. *Proc Natl Acad Sci USA* 80: 1683–1687

38 Steinmetz M, Minard K, Horvath S, McNicholas J, Srelinger J, Wake C, Long E, Mach B, Hood L (1982) A molecular map of the immune response region from the major histocompatibility complex of the mouse. *Nature* 300: 35–42

39 Kronenberg M, Steinmetz M, Kobori J, Kraig E, Kapp JA, Pierce CW, Sorensen CM, Suzuki G, Tada T, Hood L (1983) RNA transcripts for I-J polypeptides are apparently not encoded between the I-A and I-E subregions of the murine major histocompatibility complex. *Proc Natl Acad Sci USA* 80: 5704–5708

40 Waltenbaugh C (1981) Regulation of immune responses by I-J gene products. I. Production and characterization of anti-I-J monoclonal antibodies. *J Exp Med* 154: 1570–1583

41 Kanno M, Kobayashi S, Tokuhisa T, Takei I, Shinohara N, Taniguchi M (1981) Monoclonal antibodies that recognize the product controlled by a gene in the I-J subregion of the mouse H-2 complex. *J Exp Med* 154: 1290–1304

42 Klein J, Ikezawa Z, Nagy ZA (1985) From LDH-B to J: an involuntary trip. *Immunol Rev* 83: 61–77

See also other articles in this issue: Dorf ME, Benacerraf B (1985) I-J as a restriction element in the suppressor T cell system. *Immunol Rev* 83: 23–40

Murphy DB, Horowitz MC, Homer RJ, Flood PM (1985) Genetic, serological and functional analysis of I-J molecules. *Immunol Rev* 83: 79–103

43 Moller G (1988) Do suppressor T cells exist? *Scand J Immunol* 27: 247–250

44 Kronenberg M, Goverman J, Haars R, Malissen M, Kraig E, Phillips L, Delovitch T, Suciu-Foca N, Hood L (1985) Rearrangement and transcription of the beta-chain genes of the T-cell antigen receptor in different types of murine lymphocytes. *Nature* 313: 647–653
45 Eichmann K (1988) Suppression needs a new hypothesis, an answer to Göran Möller. *Scand J Immunol* 28: 273
46 Bloom BR, Salgame P, Diamond B (1992) Revisiting and revising suppressor T cells. *Immunol Today* 13: 13113–13116
47 Shevach EM (2000) Suppressor T cells: Rebirth, function and homeostasis. *Curr Biol* 10: R572–R575
48 Nishizuka Y, Sakakura T (1969) Thymus and reproduction: sex-linked dysgenesia of the gonad after neonatal thymectomy in mice. *Science* 166: 753–755
49 Sakaguchi S (2000) Regulatory T cells: key controllers of immunologic self-tolerance. *Cell* 101: 455–458
50 Taams LS, Akbar AN, Wauben MHM (eds) (2005) *Regulatory T cells in inflammation*. Birkhäuser Verlag, Basel, Boston, Berlin

Further Reading

Keating P, Cambrosio A (1997) Helpers and suppressors: on fictional characters in immunology. *J Hist Biol* 30: 381–396
Sercarz E, Oki A, Gammon G (1989) Central *versus* peripheral tolerance: clonal inactivation *versus* suppressor T cells, the second half of the 'Thirty Years War'. *Immunol* Suppl 2: 9–14
Ward K, Cantor H, Nisonoff A (1978) Analysis of the cellular basis of idiotype-specific suppression. *J Immunol* 120: 2016–2019
Weinberger JZ, Germain RN, Benacerraf B, Dorf ME (1980) Hapten-specific T cell responses to 4-hydroxy-3-nitrophenyl acetyl. V. Role of idiotypes in the suppressor pathway. *J Exp Med* 152: 161–169
Benacerraf B, Germain RN (1981) A single major pathway of T-lymphocyte interactions in antigen-specific immune suppression. *Scand J Immunol* 13: 1–10
Dorf ME, Benacerraf B (1984) Suppressor cells and immunoregulation. *Annu Rev Immunol* 127–157
Kontiainen S, Feldmann M (1979) Structural characteristics of antigen-specific suppressor factors: definition of 'constant' region and 'variable' region determinants. *Thymus* 1: 59–79

Chapter 12

References

1 Jerne NK (1985) The generative grammar of the immune system. *EMBO J* 4: 847–852
2 Chomsky N (1965) *Aspects of the Theory of Syntax*. MIT Press, Cambridge, Mass
3 Jerne NK (1984) Idiotypic networks and other preconceived ideas. *Immunol Rev* 79: 5–24
4 Schöffer N (1981) *La Théorie des Miroirs*. Ed. Belfond, Paris

References and further reading, Part II

5 Coutinho A (1980) The self-nonself discrimination and the nature and acquisition of the antibody repertoire. *Ann Immunol (Paris)* 131D: 235–253
6 Coutinho A, Forni L, Holmberg D, Ivars F, Vaz N (1984) From an antigen-centered, clonal perspective of immune responses to an organism-centered, network perspective of autonomous activity in a self-referential immune system. *Immunol Rev* 79: 151–168
7 Bretscher P, Cohn M (1970) A theory of self-nonself discrimination. *Science* 169: 1042
8 Langman RE, Cohn M (1984) The 'complete' idiotypic network is an absurd immune system. *Immunol Today* 7: 100
9 Ehrlich P (1901) Die Schutzstoffe des Blutes. 73. Vers. Deutsch. Naturforsch. u. Ärzte. *Dtsch Med Wochenschr* 50/52
10 Ramos GR, Vaz NM, Saalfeld K (2006) Wings for flying, lymphocytes for defense, exaptation and specific immunity. *Complexus* 3: 211
11 Gould SJ, Lewontin RC (1979) The spandrels of San Marco and the Panglossian paradigm: a critique of the adaptationist programme. *Proc R Soc Lond B Biol Sci* 205: 581–598
12 Vaz NM, Varela FJ (1978) Self and non-sense: an organism-centered approach to immunology. *Med Hypotheses* 4: 231–267
13 Varela FJ (1995) The Emergent Self. In: Brockman J (ed): *The third culture: Beyond the scientific revolution.* Simon & Schuster, New York
14 Maturana H, Varela F (1980) *Autopoiesis and Cognition: The Realization of the Living.* D. Reidel, Boston
15 Calenbuhr V, Bersini H, Stewart J, Varela FJ (1995) Natural tolerance in a simple immune network. *J Theor Biol* 177: 199–213
   Stewart J, Varela FJ (1991) Morphogenesis in shape-space. Elementary metadynamics in a model of the immune network. *J Theor Biol* 153: 477–498
   Varela FJ, Coutinho A (1991) Second generation immune networks. *Immunol Today* 12: 159–166
   Varela FJ, Coutinho A (1989) Immune networks: getting on to the real thing. *Res Immunol* 140: 837–845
16 Eco U (1976) *A Theory of Semiotics.* Bloomington: Indiana University Press
   Eco U (1984) *Semiotics and the Philosophy of Language.* McMillan Press, Houndsmill
17 Peirce CS, Hartshorne C, Weiss P (1960) *Collected Papers of CS Peirce.* Belknap Press
   Sassure DF (1967) *Cours de Linguistique general.* Harrassourtz, Wiesbaden;
   Sebeok TA (2001) *Global Semiotics.* Indiana University Press
18 Von Uexküll T (1999) The relationship between semiotics and mechanical models of explanation in the life sciences. *Biosemiotica* 127: 647–655
19 Celada F, Mitchison A, Sercarz EE, Tada T (eds) (1988) *The Semiotics of Cellular Communication in the Immune System.* Springer-Verlag, Berlin and Heidelberg
   See articles by Eco U, Von Uexküll T, Violi P, Golub ES, Bona CA, Ohno S, Varela FJ, Jaquemart F and Coutinho A, Vaz NM, Hoffmann GW, Sercarz EE

Further reading

Hauser A (1984) *Der Manierismus. Die Krise der Renaissance und der Ursprung der modernen Kunst.* C.H. Beck, München

Maturana HR, Varela FJ (1987) *The tree of knowledge: The biological roots of human understanding.* Shambhala, Boston

Chapter 13

References

1 Hozumi N, Tonegawa S (1976) Evidence for somatic rearrangement of immunoglobulin genes coding for variable and constant regions. *Proc Natl Acad Sci USA* 73: 3628–3632

2 Tonegawa S, Maxam AM, Tizard R, Bernard O, Gilbert W (1978) Sequence of a mouse germ-line gene for a variable region of an immunoglobulin light chain. *Proc Natl Acad Sci USA* 75: 1485–1489

3 Schlissel MS (2003) Regulating antigen-receptor gene assembly. *Nat Rev Immunol* 3: 890–899

4 Rusconi S, Kohler G (1985) Transmission and expression of a specific pair of rearranged immunoglobulin mu and kappa genes in a transgenic mouse line. *Nature* 314: 330–334
   Ritchie KA, Brinster RL, Storb U (1984) Allelic exclusion and control of endogenous immunoglobulin gene rearrangement in kappa transgenic mice. *Nature* 312: 517–520
   Baltimore D, Grosschedl R, Weaver D, Costantini F, Imanishi-Kari T (1985) Studies of immunodifferentiation using transgenic mice. *Cold Spring Harb Symp Quant Biol* 50: 417–420

5 Zijlstra M, Bix M, Simister NE, Loring JM, Raulet DH, Jaenisch R (1990) Beta 2-microglobulin deficient mice lack $CD4^-8^+$ cytolytic T cells. *Nature* 344: 742–746

6 Goodnow CC (1989) Cellular mechanisms of self-tolerance. *Curr Opin Immunol* 2: 226–236

7 Kisielow P, Bluthmann H, Staerz UD, Steinmetz M, von Boehmer H (1988) Tolerance in T-cell-receptor transgenic mice involves deletion of nonmature CD4+8+ thymocytes. *Nature* 333: 742–746

8 Schimpl A, Wecker E (1972) Replacement of T-cell function by a T-cell product. *Nat New Biol* 237: 15–17

9 Taniguchi T, Matsui H, Fujita T, Takaoka C, Kashima N, Yoshimoto R, Hamuro J (1983) Structure and expression of a cloned cDNA for human interleukin-2. *Nature* 302: 305–310

10 Mosmann TR, Cherwinski H, Bond MW, Giedlin MA, Coffman RL (1986) Two types of murine helper T cell clone. I. Definition according to profiles of lymphokine activities and secreted proteins. *J Immunol* 136: 2348–2357

11 Bjorkman PJ, Saper MA, Samraoui B, Bennett WS, Strominger JL, Wiley DC (1987) Structure of the human class I histocompatibility antigen, HLA-A2. *Nature* 329: 506–512

12 Babbitt BP, Allen PM, Matsueda G, Haber E, Unanue ER (1985) Binding of immunogenic peptides to Ia histocompatibility molecules. *Nature* 317: 359–361

13 Falk K, Rotzschke O, Rammensee HG (1990) Cellular peptide composition governed by major histocompatibility complex class I molecules. *Nature* 348: 248–251

References and further reading, Part II

14 Rotzschke O, Falk K, Deres K, Schild H, Norda M, Metzger J, Jung G, Rammensee HG (1990) Isolation and analysis of naturally processed viral peptides as recognized by cytotoxic T cells. *Nature* 348: 252–254

15 Kisielow P, Teh HS, Bluthmann H, von Boehmer H (1988) Positive selection of antigen-specific T cells in thymus by restricting MHC molecules. *Nature* 335: 730–733

16 Matzinger P (1994) Tolerance, danger, and the extended family. *Annu Rev Immunol* 12: 991–1045

17 Janeway CA Jr (1993) How the immune system recognizes invaders. *Sci Am* 269: 72–79

18 Janeway CA Jr (1992) The immune system evolved to discriminate infectious nonself from noninfectious self. *Immunol Today* 13: 11–6

19 Gay NJ, Keith FJ (1991) Drosophila Toll and IL-1 receptor. *Nature* 351: 355–356

20 Kopp EB, Medzhitov R (1999) The Toll-receptor family and control of innate immunity. *Curr Opin Immunol* 11: 13–18

21 Joao C (2007) Immunoglobulin is a highly diverse self-molecule that improves cellular diversity and function during immune reconstitution. *Med Hypotheses* 68: 158–161

22 Martinez-A C, Pereira P, de la Hera A, Bandeira A, Marquez C, Coutinho A (1986) The basis for major histocompatibility complex (MHC) and immunoglobulin gene control of helper T cell idiotopes. *Eur J Immunol* 16: 417–422

23 Eichmann K, Rajewsky K (1975) Induction of T and B cell immunity by anti-idiotypic antibody. *Eur J Immunol* 5: 661–666

24 Caton M, Diamond B (2003) Using peptide mimetopes to elucidate anti-polysaccharide and anti-nucleic acid humoral responses. *Cell Mol Biol (Noisy-le-grand)* 49: 255–262

25 Harris SL, Park MK, Nahm MH, Diamond B (2000) Peptide mimic of phosphorylcholine, a dominant epitope found on *Streptococcus pneumoniae*. *Infect Immun* 68: 5778–5784

26 Pride MW, Shi H, Anchin JM, Linthicum DS, LoVerde PT, Thakur A, Thanavala Y (1992) Molecular mimicry of hepatitis B surface antigen by an anti-idiotype-derived synthetic peptide. *Proc Natl Acad Sci USA* 89: 11900–11904

27 Sutor GC, Dreikhausen U, Vahning U, Jurkiewicz E, Hunsmann G, Lundin K, Schedel I (1992) Neutralization of HIV-1 by anti-idiotypes to monoclonal anti-CD4. Potential for idiotype immunization against HIV. *J Immunol* 149: 1452–1461

28 Dalgleish AG (1991) An anti-idiotype vaccine for AIDS based on the HIV receptor. *Ann Ist Super Sanita* 27: 27–31

29 Melnick JL (1989) Virus vaccines: principles and prospects. *Bull World Health Organ* 67: 105–112
Hiernaux JR (1988) Idiotypic vaccines and infectious diseases. *Infect Immun* 56: 1407–1413
Eichmann K, Emmrich F, Kaufmann SH (1987) Idiotypic vaccinations: consideration towards a practical application. *Crit Rev Immunol* 7: 193–227
Finberg RW, Ertl H (1987) The use of antiidiotypic antibodies as vaccines against infectious agents. *Crit Rev Immunol* 7: 269–284

Kennedy RC, Chanh TC (1988) Perspectives on developing anti-idiotype-based vaccines for controlling HIV infection. *AIDS* 2 Suppl 1: 119–127

Attanasio R, Kennedy RC (1990) Idiotypic cascades associated with the CD4-HIV gp120 interaction: principles for idiotype-based vaccines. *Int Rev Immunol* 7: 109–119

30   Lynch RG, Graff RJ, Sirisinha S, Simms ES, Eisen HN (1972) Myeloma proteins as tumor-specific transplantation antigens. *Proc Natl Acad Sci USA* 69: 1540–1544

31   Miller RA, Maloney DG, Warnke R, Levy R (1982) Treatment of B-cell lymphoma with monoclonal anti-idiotype antibody. *N Engl J Med* 306: 517–522

32   Inogès S, Rodríguez-Calvillo M, Zabalegui N, López-Díaz de Cerio A, Villanueva H, Soria E et al (2006) Clinical benefit associated with idiotypic vaccination in patients with follicular lymphoma. *J Natl Cancer Inst* 98: 1292–1301

33   de Cerio AL, Zabalegui N, Rodriguez-Calvillo M, Inoges S, Bendandi M (2007) Anti-idiotype antibodies in cancer treatment. *Oncogene* 26: 3594–3602

34   Longo DL (2006) Idiotype vaccination in follicular lymphoma: knocking on the doorway to cure. *J Natl Cancer Inst* 98: 1263–1265

35   Hurvitz SA, Timmerman JM (2005) Current status of therapeutic vaccines for non-Hodgkin's lymphoma. *Curr Opin Oncol* 17: 432–440

Timmerman JM (2004) Therapeutic idiotype vaccines for non-Hodgkin's lymphoma. *Adv Pharmacol* 51: 271–293

Coscia M, Kwak LW (2004) Therapeutic idiotype vaccines in B lymphoproliferative diseases. *Expert Opin Biol Ther* 4: 959–963

Lee ST, Jiang YF, Park KU, Woo AF, Neelapu SS (2007) BiovaxID: a personalized therapeutic cancer vaccine for non-Hodgkin's lymphoma. *Expert Opin Biol Ther* 7: 113–122

Stevenson FK, Zhu D, Rice J (2001) New strategies for vaccination and imunomodulation in NHL. *Ann Hematol* 80 Suppl 3: B132–B134

36   Bogen B, Ruffini PA, Corthay A, Fredriksen AB, Froyland M, Lundin K, Rosjo E, Thompson K, Massaia M (2006) Idiotype-specific immunotherapy in multiple myeloma: suggestions for future directions of research. *Haematologica* 91: 941–948

Houet L, Veelken H (2006) Active immunotherapy of multiple myeloma. *Eur J Cancer* 42: 1653–1660

Chong G, Bhatnagar A, Cunningham D, Cosgriff TM, Harper PG, Steward W, Bridgewater J, Moore M, Cassidy J, Coleman R et al. (2006) Phase III trial of 5-fluorouracil and leucovorin plus either 3H1 anti-idiotype monoclonal antibody or placebo in patients with advanced colorectal cancer. *Ann Oncol* 17: 437–442

Manjili MH (2007) Come forth 1E10 anti-idiotype vaccine: delivering the promise to immunotherapy of small cell lung cancer. *Cancer Biol Ther* 6: 151–152

Rhee F (2007) Idiotype vaccination strategies in myeloma: how to overcome a dysfunctional immune system. *Clin Cancer Res* 13: 1353–1355

37   Shoenfeld Y, Amital H, Ferrone S, Kennedy RC (1994) Anti-idiotypes and their application under autoimmune, neoplastic, and infectious conditions. *Int Arch Allergy Immunol* 105: 211–223

Shoenfeld Y (1994) Idiotypic induction of autoimmunity: a new aspect of the idiotypic network. *FASEB J* 8: 1296–1301

Schwartz RS (1992) Therapeutic clonotypic vaccines. *N Engl J Med* 327: 1236–1237

38 Lohse AW, Cohen IR (1991) Mechanisms of resistance to autoimmune disease induced by T-cell vaccination. *Autoimmunity* 9: 119–121

39 Weathington NM, Blalock JE (2003) Rational design of peptide vaccines for autoimmune disease: harnessing molecular recognition to fix a broken network. *Expert Rev Vaccines* 2: 61–73

40 Payne AS, Siegel DL, Stanley JR (2007) Targeting pemphigus autoantibodies through their heavy-chain variable region genes. *J Invest Dermatol* 127: 1681–1691

41 Taams LS, Akbar AN, Wauben MHW (eds) (2005) *Regulatory T cells in Inflammation*. Birkhäuser, Basel

42 Imbach P, Barandun S, d'Apuzzo V, Baumgartner C, Hirt A, Morell A, Rossi E, Schoni M, Vest M, Wagner HP (1981) High-dose intravenous gammaglobulin for idiopathic thrombocytopenic purpura in childhood. *Lancet* 1: 1228–1231

43 Fehr J, Hofmann V, Kappeler U (1982) Transient reversal of thrombocytopenia in idiopathic thrombocytopenic purpura by high-dose intravenous gamma globulin. *N Engl J Med* 306: 1254–1258

Bussel JB, Kimberly RP, Inman RD, Schulman I, Cunningham-Rundles C, Cheung N, Smithwick EM, O'Malley J, Barandun S, Hilgartner MW (1983) Intravenous gammaglobulin treatment of chronic idiopathic thrombocytopenic purpura. *Blood* 62: 480–486

44 Negi VS, Elluru S, Siberil S, Graff-Dubois S, Mouthon L, Kazatchkine MD, Lacroix-Desmazes S, Bayry J, Kaveri SV (2007) Intravenous immunoglobulin: an update on the clinical use and mechanisms of action. *J Clin Immunol* 27: 233–245

Bayary J, Dasgupta S, Misra N, Ephrem A, Van Huyen JP, Delignat S, Hassan G, Caligiuri G, Nicoletti A, Lacroix-Desmazes S, Kazatchkine MD, Kaveri S (2006) Intravenous immunoglobulin in autoimmune disorders: an insight into the immunoregulatory mechanisms. *Int Immunopharmacol* 6: 528–534

45 Dietrich G, Kaveri SV, Kazatchkine MD (1992) A V region-connected autoreactive subfraction of normal human serum immunoglobulin G. *Eur J Immunol* 22: 1701–1706

Dietrich G, Algiman M, Sultan Y, Nydegger UE, Kazatchkine MD (1992) Origin of anti-idiotypic activity against anti-factor VIII autoantibodies in pools of normal human immunoglobulin G (IVIg). *Blood* 79: 2946–2951

46 Dietrich G, Varela FJ, Hurez V, Bouanani M, Kazatchkine MD (1993) Selection of the expressed B cell repertoire by infusion of normal immunoglobulin G in a patient with autoimmune thyroiditis. *Eur J Immunol* 23: 2945–2950

47 Nimmerjahn F, Ravetch JV (2007) The antiinflammatory activity of IgG: the intravenous IgG paradox. *J Exp Med* 204: 11–15

Further reading

Kohler H, Bhattacharya-Chatterjee M, Muller S, Foon KA (1995) Idiotype manipulation in disease management. *Adv Exp Med Biol* 383: 117–122

Su S, Ward MM, Apicella MA, Ward RE (1992) A nontoxic, idiotope vaccine against gram-negative bacterial infections. *J Immunol* 148: 234–238

Bhattacharya-Chatterjee M, Foon KA (1998) Anti-idiotype antibody vaccine therapies of cancer. *Cancer Treat Res* 94: 51–68

Bendandi M (2000) Anti-idiotype vaccines for human follicular lymphoma. *Leukemia* 14: 1333–1339

Kofler DM, Mayr C, Wendtner CM (2006) Current status of immunotherapy in B cell malignancies. *Curr Drug Targets* 7: 1371–1374

Lopez-Requena A, Mateo De Acosta C, Vazquez AM, Perez R (2007) Immunogenicity of autologous immunoglobulins: principles and practices. *Mol Immunol* 44: 3076–3082

Lacroix-Desmazes S, Mouthon L, Spalter SH, Kaveri S, Kazatchkine MD (1996) Immunoglobulins and the regulation of autoimmunity through the immune network. *Clin Exp Rheumatol* 14 Suppl 15: 9–15

Chapter 14

References

1   Bona CA, Goldberg B, Rubinstein LJ (1984) Regulatory idiotopes, parallel sets and internal image of the antigen within the polyfructosan-A48 idiotypic network. *Ann Immunol (Paris)* 135C: 107–115

2   Hiernaux J, Bona C, Baker PJ (1981) Neonatal treatment with low doses of anti-idiotypic antibody leads to the expression of a silent clone. *J Exp Med* 153: 1004–1008

3   Fields BA, Goldbaum FA, Ysern X, Poljak RJ, Mariuzza RA (1995) Molecular basis of antigen mimicry by an anti-idiotope. *Nature* 374: 739–742

4   Oudin J, Cazenave PA (1971) Similar idiotypic specificities in immunoglobulin fractions with different antibody functions or even without detectable antibody function. *Proc Nat Acad Sci USA* 68: 2616

5   Jerne NK, Roland J, Cazenave PA (1982) Recurrent idiotopes and internal images. *EMBO J* 1: 243–247

6   Herzenberg LA, Chan EL, Ravitch MM, Riblet RJ, Herzenberg LA (1973) Active suppression of immunoglobulin allotype synthesis. 3. Identification of T cells as responsible for suppression by cells from spleen, thymus, lymph node, and bone marrow. *J Exp Med* 137: 1311–1324

7   Eichmann K, Coutinho A, Melchers F (1977) Absolute frequencies of lipopolysaccharide-reactive B cells producing A5A idiotype in unprimed, streptococcal A carbohydrate-primed, anti-A5A idiotype-sensitized and anti-A5A idiotype-suppressed A/J mice. *J Exp Med* 146: 1436–1449

8   Forni L, Coutinho A, Köhler G, Jerne NK (1980) IgM antibodies induce the production of antibodies of the same specificity. *Proc Natl Acad Sci USA* 77: 1125–1128

9   Coutinho A, Forni L, Holmberg D, Ivars F, Vaz N (1984) From an antigen-centered, clonal perspective of immune responses to an organism-centered, network perspective of autonomous activity in a self-referential immune system. *Immunol Rev* 79: 151–168

10 Varela FJ, Coutinho A (1991) Second generation immune networks. *Immunol Today* 12: 159–166

11 Germain RN, Benacerraf B (1980) Helper and suppressor T cell factors. *Springer Semin Immunopathol* 3: 93–127

References and further reading, Part II

12   Dietz MH, Sy MS, Benacerraf B, Nisonoff A, Greene MI, Germain RN (1981) Antigen- and receptor-driven regulatory mechanisms. VII. H-2-restricted anti-idiotypic suppressor factor from efferent suppressor T cells. *J Exp Med* 153: 450–463

13   Germain RN (2008) Special regulatory T-cell review: A rose by any other name: from suppressor T cells to Tregs, approbation to unbridled enthusiasm. *Immunology* 123: 20–27

14.   Cosenza H, Kohler H (1972) Specific suppression of the antibody response by antibodies to receptors. *Proc Nat Acad Sci USA* 69: 1701

15   Kluskens L, Köhler H (1974) Regulation of immune response by autogenous antibody against receptor. *Proc Nat Acad Sci USA* 71: 5083

16   Cosenza H (1976) Detection of anti-idiotype reactive cells in the response to phosphorylcholine. *Eur J Immunol* 6: 114–116

17   Bona C, Paul WE (1979) Cellular basis of regulation of expression of idiotype. I. T-suppressor cells specific for MOPC 460 idiotype regulate the expression of cells secreting anti-TNP antibodies bearing 460 idiotype. *J Exp Med* 149: 592–600

18   Gershon RK, Kondo K (1971) Infectious immunological tolerance. *Immunology* 21: 903–914

19   Hämmerling GJ, Black SJ, Berek C, Eichmann K, Rajewsky K (1976) Idiotypic analysis of lymphocytes *in vitro*. II. Genetic control of T-helper cell responsiveness to anti-idiotypic antibody. *J Exp Med* 143: 861–869

20   Joao C (2007) Immunoglobulin is a highly diverse self-molecule that improves cellular diversity and function during immune reconstitution. *Med Hypotheses* 68: 158–161

21   Eichmann K, Rajewsky K (1975) Induction of T and B cell immunity by anti-idiotypic antibody. *Eur J Immunol* 5: 661–666

22   Krawinkel U, Cramer M, Imanishi-Kari T, Jack RS, Rajewsky K, Makela O (1977) Isolated hapten-binding receptors of sensitized lymphocytes. I. Receptors from nylon wool-enriched mouse T lymphocytes lack serological markers of immunoglobulin constant domains but express heavy chain variable portions. *Eur J Immunol* 7: 566–573

23   Eichmann K (1975) Idiotype suppression. II. Amplification of a suppressor T cell with anti-idiotypic activity. *Eur J Immunol* 5: 511–517

24   Sachs DH (1988) Anti-idiotype to MHC receptors – a possible route to specific transplantation tolerance? *Int Rev Immunol* 3: 313–321

25   Pisetsky DS, Sachs DH (1977) The genetic control of the immune response to staphylococcal nuclease VI. Recombination between genes determining the A/J anti-nuclease idiotypes and the heavy chain allotype locus. *J Exp Med* 146: 1603–1612

26   Binz H, Wigzell H, Bazin H (1976) T-cell idiotypes are linked to immunoglobulin heavy chain genes. *Nature* 264: 639–642

27   Green DR, Webb DR (1993) Saying the 'S' word in public. *Immunology Today* 14: 523

28   Harvey MA, Adorini L, Miller A, Sercarz EE (1979) Lysozyme-induced T-suppressor cells and antibodies have a predominant idiotype. *Nature* 281: 594–596

29   Krzych U, Nanda N, Sercarz E (1989) Specificity and interactions of CD8+ T suppressor cells. *Res Immunol* 140: 302–307; discussion 339–345

30   Kumar V, Sercarz E (1993) T cell regulatory circuitry: antigen-specific and

TCR-idiopeptide-specific T cell interactions in EAE. *Int Rev Immunol* 9: 287–297

31 Urbain J (1974) Cellular recognition and evolution. *Arch Biol (Liege)* 85: 139–150

32 Brait M, Vansanten G, Van Acker A, De Trez C, Luko CM, Wuilmart C, Leo O, Miller R, Riblet R, Urbain J (2002) In: M Zanetti and JD Capra (eds): Positive selection of B cell repertoire, idiotypic networks and immunological memory. *The Antibodies* Vol 7, Taylor and Francis. p1–27

33 João C, Ogle BM, Gay-Rabinstein C, Platt JL, Cascalho M. (2004) B cell-dependent TCR diversification. *J Immunol* 172: 4709–16

34 Goldner JA (1998) The unending saga of legal controls over scientific misconduct: A clash of cultures needing resolution. *Am J Law Med* 24: 293–343

35 Ismaili J, Brait M, Leo O, Urbain J (1995) Assessment of a functional role of auto-anti-idiotypes in idiotype dominance. *Eur J Immunol* 25: 830–837

36 Binz H, Wigzell H (1977) Antigen-binding, idiotypic receptors from T lymphocytes: an analysis of their biochemistry, genetics, and use as immunogens to produce specific immune tolerance. *Cold Spring Harb Symp Quant Biol* 41: 275–284

Part III
Science between Fact and Fiction

Chapter 15

The fictional nature of scientific notions

> ....*scientific practice not only leads to the material transformation of fictional entities into real ones. It also results in the downgrading of the latter to the former.*
> P. Keating, A. Cambrosio (1997) Helpers and Suppressors: On Fictional Characters in Immunology. *J Hist Biol* 30: 381

Popper[1], Kuhn[2], and Fleck[3], as diverse as their science philosophies might appear (see chapters 3, 4), have agreed in noting that all hypotheses, paradigms, thought styles, respectively, have been transient episodes in the history of science, notions replaced by new notions soon to be revised again. Notions* in science may be based on experimental observations that may be fundamentally correct. Nevertheless, the experimental observations acquire their meaning only if they are interpreted. In order to make scientific sense, observations are explained within the constraints of a current thought style. What results are fictional notions of imagined realities.

Only very few scientists have realized, and even fewer have admitted to, the erratic nature of the scientific method. One of the few was Peter Medawar, together with Macfarlane Burnet bearer of the 1960 Nobel Prize for Physiology or Medicine, for the discovery of acquired immunological tolerance. Medawar lectured to the American Philosophical Society[4]: "Methodologists who have no personal experience of scientific research have been gravely handicapped by their failure to realize that nearly all scientific research leads nowhere – or if it does lead somewhere, then not in

---

* The term *notion* is used here in a more general meaning than terms such as paradigm, thought style, etc., which denote the body of learned knowledge that guides, though transiently, scientific practice in a given field. The term *notion*, in addition, includes all types of conclusions, interpretations, concepts, etc., that scientists or laboratory groups derive from their practical or theoretical work, from the small everyday hypotheses to comprehensive theories.

the direction it started off with. In retrospect, we tend to forget the errors, so that 'The Scientific Method' appears very much more powerful than it really is, particularly when it is presented to the public in the terminology of breakthroughs, and to fellow scientists with the studied hypocrisy expected of a contribution to a learned journal. I reckon that … about four-fifth of my time has been wasted, and I believe this is the common lot of people who are not merely playing follow-my-leader in research…. The process by which we come to formulate a hypothesis is not illogical, but non-logical, i.e. outside logic. But once we have formed an opinion, we can expose it to criticism, usually by experimentation; this episode lies within and makes use of logic". While I agree with Arthur Silverstein[5] that the final sentence of the quote seems somewhat overoptimistic (see chapter 18), I still feel that Medawar's insight deserves the highest respect.

And yet, as I have elaborated in chapter 2, science generates robust knowledge, information about the world around us that remains valid outside the laboratory and enables us to manipulate nature in such a way that predictable changes are introduced, for better or worse, but reliable and reproducible. This is not to say that all consequences of our manipulations have been predicted. They have not, owing to limitations in our knowledge about downstream effects in a complex network such as nature. Unpredicted consequences, however, do not justify arguments that negate the validity of a solid, though incomplete, body of knowledge about nature that has emerged from scientific methodology.

In the following chapters I am going to argue that robust knowledge about nature is a fortuitous but predictable byproduct of scientific methodology. Initially unintended and uncertain fragments of information develop into harder and harder facts by frequent repetition, give rise to a meaningful set of connections with previous and subsequent knowledge, and eventually get translated into procedures, products, or apparatus suitable for the manipulation of nature. However, and this is important, of the many fragments of information produced by scientists, only very few develop into robust facts. The four-fifth of Medawar's wasted time may be a fair estimate for a brilliant scientist, but overall the proportion of waste in science is certainly greater. The vast majority of observations made by scientists are not repeated, do not form meaningful connections with existing knowledge, and will never give rise to successful practical procedures. Most scientists at most times produce results that give rise to inconsequential notions, if they give rise to notions at all. And most importantly, it is impossible to distinguish such notions, *in statu nascendi*, from the few that will eventually add to our solid body of knowledge. Thus, while all notions in science are fictional, very few turn from fiction to fact and develop into robust knowledge.

The network theory is an example of a fictional notion in science that did not develop into robust knowledge. However, unlike the vast majority of inconsequential notions which are forgotten sooner rather than later after

The fictional nature of scientific notions

being generated, the network theory enjoyed a period of smashing success, it developed into a Kuhnian paradigm that governed the activities of the better part of a generation of immunologists. The network theory is thus a fictional notion under a magnifying glass, an enlarged specimen of what scientists produce all the time. It can thus serve as a paradigma in science theory, telling us how fictional notions are generated, may even enjoy a period of blossom, but eventually remain what they have always been, fiction.

Most experimental scientists are guided by the conviction that the results produced in their laboratories reveal facts, i.e. constrain the number of possible ways in which a certain sector of nature functions. An initial observation is followed by chains of experiments designed to choose among, and further narrow down, an initially large number of possible explanations, until the evidence appears to unequivocally support a single interpretation. Then they publish a paper, i.e. a text in which a description of the chain of results is followed by a concluding statement on how the sector of nature under study likely functions. The reason why such statements are always fictional notions is that, in order to develop them, the scientist unconsciously relies on a body of learned knowledge on the subject, the "theoretical background", composed of some robust facts but also of many items and connections that, unknown to the scientist, have never been unequivocally proven. The learned body of knowledge constrains the scientist's ability to see the full spectrum of possibilities for interpreting the chain of results, which in reality is very much larger. Moreover, the learned body of knowledge, whether it is termed paradigm, thought style, web of belief, or else, is unstable, it changes in time. Whether or not the statement made in the paper remains a fictional notion or develops into a robust fact depends on the unforeseeable results of future trials of strength in other laboratories and in the real world (see chapter 2, and below).

That most of what scientists produce is fiction is not a problem, it rather is an inherent property of scientific methodology and a prerequisite of its overall success. To interpret their experimental observations, scientists utilize a combination of learned knowledge and imagination. If the use of imagination were forbidden in science, no progress would be possible. As a matter of fact, scientific methodology has proven over and over again that some of the produced fiction turns into robust knowledge, some of it even valid in the real world. The process is stochastic, unpredictable for the individual instance, but occurs with a given predictable frequency, comparable to mutations in the genome. Just as mutations in the genome are the basis of the evolution of species, stochastic events that turn scientific fiction into fact are the basis of the evolution of knowledge.

What is more of a problem is that scientists are not conscious of the fictional nature of their notions. On the contrary, if a result is repeatable and seems to "make sense", the notion derived from it is usually published as definitive, and more often than not later exposed as exaggerated, inappropriately generalized, only partially correct, or even totally erroneous. More seriously, multiple instances are known, for example in medicine, in which research results have been launched in the public media combined with promises of cure, only to be later exposed as premature, overoptimistic, or even unrealistic. The science lobbyists who favor this type of strategy are the same who continually lament about the poor acceptance of science in the general public. As a result, the general public is more conscious of the fictional nature of scientific notions than the scientists are themselves.

There are multiple reasons for this partial blindness in science, and some of them will be discussed here. The origins of this ignorance already begin in education. From the beginning of their studies, students are told that one of the most important properties for a scientist to develop is an uncompromising scrutiny of all observations and opinions encountered in the professional context. Interpretations of experimental observations are to exclude all subjective criteria and personal preferences, only to be guided by pure logical reasoning. In reality, scientific education is quite different. A young scientist is trained to believe in what is being taught in textbooks and lectures, a host of information which is impossible to question by logic or to otherwise critically evaluate. Ludwik Fleck used the term "initiation" for this process: "Any didactic introduction to a field of knowledge passes though a period during which dogmatic teaching is dominant. An intellect is prepared for a given field; it is received into a self-contained world and, as it were, initiated." Thus, not unlike a novice on initiation, a junior scientist is asked to believe in a body of authoritative knowledge which has to be accepted as dogma. Because textbook knowledge cannot be scrutinized, a student can either accept or reject it. Only those prepared to accept can enter the esoteric realms of science.

The education of a scientist starts with textbooks and lectures but does not end there. A postdoctoral fellow in the laboratory is exposed to an array of existing notions that define the field of study for the laboratory. In contrast to the textbook knowledge, the notions in a laboratory very often are not generally accepted, they are controversial and sometimes matters of disagreement among different laboratories. Such laboratory-specific notions are often determined by the head of the laboratory, and a junior member of the laboratory usually accepts such notions which, although not undisputed, are well supported by evidence. That a competing notion may be equally well supported by evidence is either unconsciously ignored or actively suppressed. Thus, important elements in advanced scientific education are the convincing power and the professional authority of the teacher. The head of the laboratory often is part of a circle of colleagues sharing a set of notions, into which the junior member is gradually introduced. Introduction and acceptance in this circle further strengthens the acceptance of the shared notions, although the fact that they are controversial means that strong evidence and good arguments exist to challenge them. The junior scientist is thus educated to make a biased choice between equally strong bodies of evidence.

The fictional nature of scientific notions 211

Some scientists will argue that authoritative learning is only required for a short initial period, and that the sooner a student develops a critical mind the better a scientist he/she will become. Once developed into a true expert, only pure reasoning will guide the scientist's mind. Fleck argues to the contrary: "...the expert is already a specially molded individual who can no longer escape the bonds of tradition and of the collective; otherwise he would not be an expert". In other words, according to Fleck, initiation and conditioning are necessary developments in the transformation of a student into an expert in a particular science. For Fleck the constraints of a thought style are a necessary prerequisite of cognition, without such constraints results derived from experiments cannot be interpreted, they remain meaningless. Few modern science theoreticians would disagree with Fleck in this very basic statement.

Of course, with time and experience a scientist may come to reject certain learned notions and may begin to prefer alternatives, either existing or novel, products of his/her own thinking or that of others. Here begins what science philosophers have referred to as "theory choice", a much discussed subject[6]. Those who favor underdetermination have proposed that, because in their opinion any theory can be reconciled with any type of evidence (see chapter 2), scientists can choose among competing theories only by non-scientific criteria. While this is certainly exaggerated, there is no doubt that notions favored by charismatic science teachers are often more successful than others. As pointed out by Knorr-Cetina[7] (chapter 4), successful science teachers spend a lot of time advertising their notions in congresses and lectures, and thus receive large numbers of applications, from which they can choose the most promising collaborators. After gaining independent reputation such collaborators, with the help of their supportive mentor, will set up their own laboratories and further spread the good gospel. Distribution of materials, such as monoclonal antibodies, nucleic acid probes, genetically manipulated mice, etc., all of this helps in multiplying the influence of a laboratory in the scientific community. Because review boards consist of peers that are subject to that influence, successful laboratories get their grants approved more often than less successful ones, resulting in increased resources. By such positive feedback, the notions favored by one laboratory come to be more successful than those of others, they may displace competing notions, become dominant in the field, and may even give rise to a paradigm.

Creating a paradigm, a thought style that governs scientific practice of an entire field, is probably the ultimate satisfaction that a scientist or laboratory can come to enjoy in science. This, however, is not possible by non-scientific factors alone. It requires that a pressing but unsolved problem, or an entire field of problems, becomes amenable to be attacked by novel ideas, assay systems, methodology, apparatus, etc., in a way that the scientific community feels that significant progress can be made. The experimental systems have to be sufficiently robust to be set up in other laboratories, with reproducible results under a variety of different basic laboratory conditions[8]. The distributed reagents have to be reliable as well, their performance independent of the investigator handling them. As a minimum, the experimental systems and notions giving rise to the paradigm have to stand the trial of strength in general laboratory conditions, outside the boundaries of the laboratory of origin.

Does this mean that notions, in order to give rise to paradigms, have to have left the realms of fiction and turned into fact? Paradigms are inherent to scientific practice and usually do not reach out into the real world. A paradigm, influential as it may be in laboratory science, can be largely fictional, as the network paradigm shows. Paradigms will eventually be discarded or replaced by new ones, even though certain components may be maintained for extended periods of time. However, occasionally such a component has survived many paradigm shifts, has been maintained ever since it has first been proposed, and continues to be maintained. It has developed into a scientific fact, is now taken as a matter of course, a basic scientific truth to be included, first outspokenly but later without mention, as a "given" in all subsequent notions in that field. Creation of a paradigm may thus be a first step in the generation of a scientific fact, but there is no guaranty. The generation of a scientific fact, however, is only the first step in the development of what I refer to as robust knowledge. On the way to become, or not to become, robust knowledge a scientific fact will be handed over to technicians, engineers, physicians, etc., who put scientific facts to the test in the real world. Moreover, judgements of social, political, ethical, and economic origin have their role in admitting scientific facts to the body of human knowledge. The generation of knowledge from scientific notions includes thus a two-tier selection process: Selection in the laboratories and selection in the real world. Only a minority survives.

Chapter 16

Fiction turned fact:
The case of antibodies

...we speak of antibodies as if they exist, while reserving judgement as to the objective reality of their existence.
Louis Hallion (1906) Specificité des anticorps:
Sensibilisatrice et alexine. *Presse Médicale* 13, 82: 653

In chapter 2 I have discussed the discovery, by Oswald Avery and colleagues, of DNA as the molecular principle of pneumococcal transformation[1] and the subsequent development of the fact, now robust knowledge, that DNA is the molecular principle of heredity in all of nature*. One can discuss at length which of the various subsequent experiments established this fact once and for all, but most would agree that the solution of the three-dimensional structure of DNA by X-ray crystallography by Watson and Crick in 1953[2] had a pivotal role, by suggesting readily comprehensible mechanisms for replication, one of the two faculties the hereditary material must possess. Others might insist that the matter was not settled until Nirenberg and Matthaei, in 1961, added the mechanism by which DNA codes for proteins[3], the second of the two essential faculties. In either case, considerable time had passed and effort had been spent since Avery's initial notion in 1944.

Just like it is commonplace today to refer to Avery and colleagues as the discoverers of the genetic function of DNA, history of immunology has it that von Behring and Kitasato discovered antibodies when they, in 1890, reported their observations on the anti-toxin activity in the sera of animals pretreated with bacterial toxins[4]. However, while this was no doubt a landmark observation, the notion of antibody took decades to develop and was not turned into fact until many years later. Indeed, one can argue that anti-

---

* For a discussion of exceptions in biology see chapter 18.

bodies did not rise to the state of scientific fact until Porter (1959)[5] and Edelmann (1961)[6,7] reported their pathbreaking studies on the subunit structure, thus all of a sudden suggesting readily comprehensible models of antibody structure and function.

In chapter 6 I have given a compressed outline of the history of the antibody paradigm, restricting myself to a selection of notions that were key to the advancement of the field, as if this had been a straightforward goal-oriented process. Here now I will concentrate on the multiple detours and vagaries, scientific but also personal and political in nature, in the history of conceptual thinking about antibodies, a field of study that exemplifies the erratic process of scientific fact finding particularly well. As pointed out by Lindenmann[8], in the two 1890 papers on the immunity to diphtheria by von Behring[9], and to diphtheria and tetanus by von Behring and Kitasato[10], the authors carefully avoided the noun 'anti-toxin', and the adjective 'antitoxic' occurs in a footnote, used in the sense of 'antiseptic'. In a paper of 1892 together with his long-term coworker Erich Wernicke, von Behring still did not use the term antitoxin[11]. Von Behring and his coworkers did not think in terms of a defined molecular entity or substance, endowed with the antitoxic function, that appeared in sera of animals as a result of their having acquired immunity to diphtheria. Rather, they entertained the notion that the antitoxic activity was a property of the serum as a whole, some activity emerging from complex alterations in its composition. Their concept was that "the disinfecting action of the serum of artificially or naturally immune animals is linked to the cell-free fluid of the blood – the serum."* What might be taken as a reference to a putative anti-toxic substance in immune serum appears first in von Behring's 1894 booklet: *Das neue Diphtherie-Mittel* (The new Diphtheria Remedy)[12], where he writes "that the healing of a specific disease is accompanied by the production of specific toxin-antagonistic (anti-toxic) agents ... that circulate in the blood". However, "we ourselves know nothing about the nature of the acting principle in the blood, and only insofar are we informed about it, that we ourselves stay away from a so-called 'purification'". Putting the purification in quotation marks clearly reveals von Behring's conviction that to the informed expert any attempt at obtaining a pure active substance would seem ludicrous. Only much later would he refer to the activity as "immune bodies" or "healing bodies"[13], the term "bodies" being in use at the time for protein components in general.

The term "antitossina" as referring to a defined substance appeared first in papers of 1891 by G. Tizzoni and G. Cattani of Bologna[14,15], who did protection experiments in animals infected with tetanus. They had immunized dogs and were the first to attempt an enrichment of the anti-toxic activity by fractionated magnesium sulfate precipitation of the dog serum. Among the

---

* As pointed out by Silverstein[4], the claim that immunity was linked to serum was a highly controversial issue as it contradicted Ilya Metchnikov's theory that immunity was mediated only by cells, i.e. macrophages and microphages.

Fiction turned fact: The case of antibodies

two fractions obtained, "albumin" and "globulin", they found the activity in the globulin fraction and reported success in protection of infected rats, a finding which in a Belgian textbook on veterinary microbiology of 1892 was praised as a pioneering advance[16]. In contrast, von Behring was not at all impressed with their results, as he usually was very much concerned with issues of priority in all aspects of serum therapy. He wrote that "their substance, extracted from serum, hardly has any reliable immunizing value, not to speak of healing value". He continues, "we should not be satisfied with the qualitative demonstration of a specific healing action, with which Tizzoni was content. His antitoxin has no immediate effect in rabbits and guinea pigs, for him a qualitative demonstration of a healing effect in rats suffices."[13]

One should not blame von Behring that he did not believe in proteins as molecular entities that possess defined functional properties. His imagination was constrained by the thinking of the late 19th century on the material basis of living matter, which was determined by two alternative, but not really mutually exclusive, schools of thought[17]. The "chemical school", widely disseminated by a leading textbook of Max Verworn[18], envisaged the cellular substance as a giant lump of "living albumin" associated with interspersed chemical compounds that formed and disassembled as long as the cell was alive. When the cell died, the interspersed chemical matter disintegrated and vanished whereas the albumin remained, as "dead albumen". The contrasting "physical school" of Thomas Graham[19] distinguished "colloids", which are mostly proteins and the basis of living matter, from "crystalloids", which are typical of the inorganic substance. The colloids, also called "bodies", and the crystalloids are distinct by their degree of diffusibility, which is low for the former and high for the latter. Colloids are gelatinous and inert, in so far as they do not ionize or engage in chemical reactions. Their chemical indifference makes the colloids uniquely suitable for the organic processes of life, they form the plastic elements of the animal body, they are highly mutable and dynamic, they possess "energia". While they do not take part in chemical reactions, colloids interact with one another and with crystalloids by non-stoichiometric processes like adsorption, layer formation, aggregation, and the like. It is obvious that these thought styles have no room for concepts that ascribe defined functional properties to isolated protein components.

The term "antibody" first came up in a paper by Paul Ehrlich of 1891, in which he was making the point that two substances must be chemically different if they elicit different antibodies[20]. It was at first not a successful term, however, as it was rarely used in the scientific literature for many years to come[8]. Immunologists stuck to the multiple technical terms, derived from the methods of detection, such as "agglutinin", "precipitin", "hemolysin", etc. There was also quite a heterogeneous collection of terms that referred to the protective effect, including "immune bodies", "immunisin", "protection bodies", "healing bodies", as well as terms referring to the putative sensitizing function for complement in hemolysis, such as "amboceptor", "intermediate body", "substance sensibilatrice", "fixateur", only to name a few in each category. The multitude of terms only documents that one could not be sure what one was talking about, nor were there any clear notions on whether such terms designated different or similar entities. Indeed, the problem was still not solved when in 1929 Harry Gideon Wells, in his book "The Chemical Aspects of Immunity", proposed the "unitarian hypothesis" according to which all antibodies, independent of the assay by which they were detected, represent a single class of molecules[21].

The initial excitement with von Behring's serum therapy threatened to fade away when it became obvious that many of the sera produced in his laboratory failed in animal protection trials, supposedly because of insufficient anti-toxic activity. Also elsewhere, for example in Britain, attempts to produce protective sera by von Behring's methods were largely unsuccessful. Von Behring himself was extremely reluctant to begin human trials, as he foresaw and feared the severe disappointment of his financiers and colleagues if initial trials would yield unsatisfactory results[13]: "We have to reckon with the unfavorable possibility and have to be prepared for the case that the outcome of the first (human) healing trials may not be unequivocally positive. Under these circumstances, and because of the reason that we ourselves abstain from performing preliminary orienting human trials, we believe it is useful to recommend to those who want to begin (human) trials with our diphtheria-remedy, not only to demonstrate the complete harmlessness in animals, but also to gain an independent judgement that the therapeutic strength of our remedy may be improved by an unforeseeable margin in the future". In other words, the situation called for some procedure of calibration, in order to be able to test animals during the immunization procedure so as to optimize injection schedules, to compare different batches of sera, to select the optimal species for immunization, to define optimal therapeutic doses, etc.

Ehrlich, who was much more than von Behring prepared to break with current thought constraints, developed an interest in using the diphtheria anti-toxins to study the quantitative aspects of antibody function. Other than most of his contemporaries, Ehrlich believed in the chemical reactivity of proteins including that of antibodies. He had studied the reactivity of dyes like methylene blue with tissue components, and had observed that these reactions were completely reversible when the stained tissues were extensively washed with solvents. In contrast, he found the binding of anti-toxins to the toxin much more resistant to dissociation, suggesting some form of chemical reaction, and stoichiometric rules to apply to toxin-anti-toxin binding. Initially by agreement with von Behring and with his financial support, Ehrlich began to study the stoichiometry of diphtheria toxin-antitoxin binding, using guinea pig lethality and protection as a quantitative assay. Very rapidly, however, von Behring complained about "the elimination of my influence on the (Ehrlich's) Frankfurt institute" and the situation

Fiction turned fact: The case of antibodies   217

developed into a scientific priority conflict, leading to a severe fallout between the two when von Behring cancelled the financial contribution he had agreed to provide for the project[13,22].

Ehrlich's disappointment is documented in a letter to his secretary and biographer, Martha Marquardt[23], in which he complained bitterly about von Behring:

> "...that the institute should be a subsidiary to (Behring's institute in) Marburg and that I should be his subordinate was never mentioned.... I can readily believe that B. has since changed his mind and imagined I would work for some derisory salary to ensure him a flow of new discoveries and extra millions....
>
> He has only me to thank for the diphtheria success and especially his great material gains. When we joined forces, he had only a one-quarter to one-half-fold serum while I had a thirty-fold one....
>
> Since I left him, he has been able to see how far he has got without me. Everything has been a failure – plaque, cholera, glanders, streptococci, no progress in the diphtheria sector, and only a risky hypothesis and pseudo-exact juggling with figures – and all that with a surfeit of funds and a swarm of workers."

The fallout was transient, however, and the two later reestablished a good working relationship combined with mutual respect and even collegial esteem[22].

While working on the calibration of diphtheria toxins and antitoxins Ehrlich discovered what some historians have referred to as the Ehrlich-phenomenon or –paradoxon[17]. When he had titrated a serum to contain X units of toxin-neutralizing activity using a fresh standard batch of toxin, other batches of toxin adjusted to equal strength as the fresh standard required mostly more units of toxin-neutralizing activity of the same serum. Soon he realized that the older the toxin preparation was, the more neutralizing units were needed to neutralize the same amount of toxicity, resulting in decreasing neutralizing units found in the same serum the older the test batch of toxin was. Ehrlich resolved this seemingly paradoxical result by postulating two different reactive groups on the toxin, one reacting with the antitoxin and relatively stable, the other endowed with the toxic activity and more labile with time. Ehrlich dubbed these the "haptophore" and the "toxophore" groups, respectively[24]. He developed a complicated scheme of progressive modification of the toxophore group and envisaged multiple stages of degeneration for which he invented terms like

---

*   Ehrlich's theory of detoxification raised some antagonism among his colleagues. For example, Svante Arrhenius wrote in 1907: "Ehrlich and his school invented the artificial hypothesis that these poisons consist really of a great number of different poisons and innocious substances which combine with antitoxin... Nearly every new phenomenon led him and his "toxon", "prototoxoid", "deuterotoxoid", "tritotoxoid", of which only the term "toxoid" survived, designating a non-toxic preparation of toxin*.

What also survived was the concept of molecules containing two or more functional groups, which remained dominant in his thinking ever since. For example, in a paper of 1901 on the hemolytic function of antibodies[25] he compared antibodies to diazobenzaldehyde, a small chemical compound in which two reactive groups are attached to a benzene ring, both of which able to form new compounds by reacting with other chemicals. For drawing the analogy to the hemolytic effect of antibodies on red blood cells, he envisaged that the diazo-group would react with a structure of the red cell while the aldehyde-group would interact with a toxic compound in the solvent, the "complement", thus concentrating complement on the cell causing its destruction.

These examples show that Ehrlich, in interpreting experimental evidence, let his imagination run freely and daringly into novel territory. Many of his fictional notions were novel and intellectually fascinating, and they seemed to open the door to the understanding of serum therapy, a conceptual void in medical science. As a result, Ehrlich became one of the leading authorities on immunity at the beginning 20th century, and was one of the very few foreigners ever invited to deliver the Croonian lecture to the Royal Society of London, which he did in March 1900. There he outlined in detail his observations on toxin modification, and then went on to discuss the side chain theory[26], previously presented in rough form in 1897[24], which made him famous and earned him the Nobel Prize in Physiology or Medicine in 1908, a perfect example of scientific fiction which, in many of its elements, was rather far off the mark.

In brief, the side chain theory postulated the existence of multiple "atomic groups" which are attached to the protoplast of each cell of the body, sticking out towards the exterior of the cell, allowing it to interact with compounds in the environment. The idea of a cell as a giant molecule with functionally reactive groups attached to it was built on the diazobenzaldehyde model of the antibody, only transformed to a larger scale. The physiological function of the side chains in cellular physiology was to capture foodstuffs, such as sugars, in the body fluids in order to be taken up by the cell as nutrients. Only by pure chance do some of these side chains interact also with the haptophore groups of toxins, presumably by some kind of molecular likeness to certain nutrients. Ehrlich argues that it is difficult to envisage biological reasons for toxin-binding groups on cells: "That they are in function especially designed to seize on toxins cannot be for one moment entertained. It would not be reasonable to suppose that there were present

---

school to invoke the presence of a new substance; owing to this circumstance the theory of Ehrlich has to a great extent lost its credibility". Arrhenius S: *Immunochemistry, the application of the principles of physical chemistry to the study of biological antibodies*. New York, 1907.

in the body many hundreds of atomic groups destined to unite with toxins, when the latter appeared, but in function really playing no part in the processes of normal life and only arbitrarily brought into relation with them by the will of the investigator. It would, indeed be highly superfluous, for example, for all our native animals to possess in their tissues atomic groups deliberately adapted to unite with abrin, ricin, and erotin, substances coming from the far distant tropics". Stunningly, in spite of the large death toll in his time due to infections like diphtheria and tetanus, Ehrlich could not envisage that the coevolution of vertebrates with toxin-producing bacteria has brought about a selection pressure on the former towards developing antitoxic resistance mechanisms. In his view, the antitoxin immune response in mammals was an irrelevant byproduct of the requirement of the cell to pick up nutrients.

Ehrlich went on to propose that the soluble antibodies in the blood were nothing other than the shed side chains. For explaining the mechanism of shedding he borrowed an idea from the scientist Carl Weigert[27], Ehrlich's cousin, who was interested in tissue regeneration. Ehrlich: "The side chain involved (in toxin binding), as long as the union lasts, cannot exercise its normal physiological nutritive function – the taking up of definite foodstuffs. We are therefore now concerned with a defect which, according to the principles so ably worked out by Professor Carl Weigert*, is repaired by regeneration. These principles, in fact, constitute the leading conception in my theory. ...the side chains, which have been reproduced by the regenerative process, are taken up again into union with the toxin, and so again the process of regeneration gives rise to the formation of fresh side chains ... the cells become, so to say, educated or trained to reproduce the necessary side chains in over-increasing quantity ... over-increasing regeneration must finally reach a stage at which such an excess of side chains is produced that ... the side chains are present in too great quantity for the cell to carry, and are, after the manner of secretion, handed over as needless ballast to the blood."

Phrases such as "highly superfluous" or "needless ballast" reveal that the side chain theory, elegant and in itself consistent as it was, envisaged the process of antibody production as a rather useless deviation of normal body physiology, essentially an artifact of deliberately injecting foreign substances into laboratory animals. Ehrlich clearly recognized the beneficial nature of antitoxic therapy for human infections, but he explained the effect as a prevention of cellular starvation, and not as a toxin-neutralizing effect. In his view passive immunization was a medical procedure to transfer protection, but he failed to see that in many other infections protective anti-

---

* Weigert's law: "Loss or destruction of tissue results in compensatory replacement and over-production of new tissue during the process of regeneration or repair (or both), as in the formation of callus when a fractured bone heals". Weigert C (1896) *Deutsche Medizinische Wochenschrift* 22: 635.

bodies were produced actively. Indeed, the aspect of active immunity against infection is completely missing in the side chain theory, and not in one sentence did Ehrlich consider the possibility that active antibody production may be an important bodily function in animals and humans in order to survive in an environment full of pathogenic microbes and their toxic products. In spite of Ehrlich's many precise experimental observations and conceptual contributions to antibody function, it is correct to say that the side chain theory, in addition to its understandable flaws in biological detail, missed the most important biological aspects of antibody production, namely the essential role in fighting infections. From Jenner to Pasteur, the experimental observations that could have served as a basis for a theory of active immunity against infections had been reported and were common knowledge at the time. Ehrlich had asked a British coworker, one Dr. Bashford, to translate his draft of the Croonian lecture so that it could be published in English. Significantly, the term 'Antikörper' was one of very few terms left untranslated and used by Ehrlich in the original German. Antibodies still had a long way to go from fiction to fact.

Early observations on antibody function were repeated in due course by the French researchers Jules Bordet[28] on antibodies that lysed red blood cells and J. Danysz[29] on antitoxins to the plant toxin ricin, and by the Austrians Philipp Eisenberg and Richard Volk[30] on antibodies to typhoid bacilli. As vividly described by Schröder-Gudehus[31], a French-German priority conflict developed, reaching its summit during World War I, in which each side insisted on the national terms for antibodies (see above) and for the heat-labile serum substance that caused hemolysis, dubbed "complement" by Ehrlich, "alexin" by Bordet[28], not to speak of several other terms in use[32,33]. Using civilized language, Ehrlich wrote[34]: "with respect to complement we could not adopt the unitarian viewpoint of Bordet, but on the basis of our experiments we arrived at the conclusion...". More aggressively, Bordet referred to Ehrlich's illustrations of the side chain theory as "quite puerile graphic representations", stating that it was "irrational" to attribute "a given property with which an antibody is endowed to particular atomic groupings inscribed in its molecule"[35]. Félix Le Dantec compared the antitoxic activity of serum with the sleep inducing effect of chloral, and ridiculed Ehrlich as "the German scholar" who postulated that chloral must therefore contain a "dormitine"[36]. However, also independent of the French-German conflict the controversies over antibodies continued for decades. Much scientific opposition arose against the notion of antibodies, up to the complete negation of their existence. Henry Dean wrote in 1917: "Agglutinins, precipitins, amboceptors, are mere words, and a passive belief in the existence of such bodies tends to impede rather than advance our understanding of what is actually taking place."[37]

In the following decades research in immunity was dominated by studies of antigenicity, whereas progress in knowledge on antibodies was slow. As pointed out by Lindenmann[8], the term 'antigen' was introduced in 1903

Fiction turned fact: The case of antibodies

by Ladislav Deutsch, aka Lásló Detre, a hungarian researcher who, when he worked with Ilya Metchnikow in Paris used, in a paper written in French, the phrase 'substances immunogènes ou antigènes'[38]. The terms resulted from a misconception because Deutsch believed in the theories of Buchner (1893)[39] and of Emmerich and Loew (1901)[40] that immunity was mediated by the bacterial toxins themselves that were modified by the host organism into antitoxins. Deutsch's 'substances antigènes' were bacterial products in transition to antibacterial products, or antigens about to turn into antibodies.

In 1912 the Austrian scientist E.P. Pick wrote a comprehensive review on the nature of antigens[41] which seemed to summarize the present knowledge: "There is one definite characteristic possessed by all antigens, and that is, that they are all colloids ... the second, which is intimately connected with the first, is that all true antigens are proteins. All attempts to raise antibodies with non-proteins have so far failed, so that one is justified in making the statement – no antigen without protein". Interestingly, as pointed out by Mazumdar[17], already at the time there was published evidence to the contrary, for example by J.J. Abel and W.W. Ford of Johns Hopkins University who in 1907 reported on antibodies to a fungus hemolysin which they had identified as a glycoside[42]. To hold on to the dogma, Pick explained this in the review by making the ad hoc hypothesis that the glycoside must have been attached to a protein, thus furnishing a great example for Quine's doctrine that any scientific theory can be reconciled with any evidence (see chapter 2). Ironically, there would have been some truth to Pick's ad hoc hypothesis had he known about helper T cells and their requirement for antibody production. However, what Pick meant by antigen was a substance that can be bound by an antibody, and in this context the claim that all antigens must be proteins was already outdated when his review was written. Nevertheless, for the case of polysaccharides it took until 1923 to settle the matter, when Michael Heidelberger and Oswald Avery identified the group-specific antigen of pneumococcus as a polysaccharide[43]. Moreover, the question of antigenicity was not settled until Karl Landsteiner, during his appointment in The Netherlands around 1920, observed that antibodies can be raised against chemical groups artificially attached to proteins, so-called haptens, suggesting that anything, even molecules that do not occur in nature, can be antigenic[44].

The history of concepts of the nature of antigens is as erratic as that of antibodies, as the above examples show. One could fill entire chapters listing further examples of experimental errors and faulty conclusions. There is no doubt that the history of research on antibodies and antigens is full of errors and misconceptions, even though in the end, as we know today, both terms are connected to a body of solid knowledge that stands any trial of strength not only in the laboratory but in the real world as well. Why then is scientific methodology so error prone? Is it perhaps so that those who committed the errors were scientists of lesser quality and competence than those who contributed the correct results? Were the notions that caused setbacks and detours in progress perhaps contributed by researchers of sub-optimal intelligence and qualification? In the remaining of this chapter I will describe a few examples of erroneous contributions of scientists whose pivotal role in advancing the field is beyond any reasonable doubt. The examples serve me to corroborate my conclusion that errors are part of normal research, both in experiments and interpretations.

The first example concerns an error committed by Karl Landsteiner, referring to Ehrlich and Julius Morgenroth who, between 1899 and 1901, had published a series of six papers[45] on hemolysins, in which they developed the concept of the antibody as "amboceptor" forming a bridge between the red cell on the one hand and the complement, the hemolytic toxin, on the other. In addition, from experiments demonstrating crossreactions of the antisera by absorption with red cells derived from species other than that whose red cells were used for immunization, they concluded that an antiserum contains a multitude of different antibodies reactive with a multitude of structures on the red cell; some of these structures are unique but others are shared by heterologous red cells which are therefore also lysed by the antiserum, though with lower activity. Ehrlich's and Morgenroth's interpretation could not have been more correct.

As a young man working in Vienna, Karl Landsteiner had studied the agglutinating activity of normal human serum for the red cells of other humans and thereby discovered the human blood groups[46], for which he received the Nobel Prize in 1930. After having moved to New York he became interested in serological crossreactions between serum proteins of different species, which he studied by sequential precipitation of antisera to whole allogeneic serum, using whole sera of different heterologous species as antigens. In 1924 he published a paper on these experiments, which also included studies on crossreactions among red blood cells of different species determined by agglutination assays[47]. In the latter, he got results similar to that of Ehrlich's and Morgenroth's absorption experiments. With respect to the former, he observed "that precipitins in general show a maximum activity with the homologous protein, the reaction decreasing gradually in strength with the distance in the zoological scale when other antigens are tested". From these observations he developed the – correct – idea that a precipitin may react with other substances if "their chemical structure is sufficiently near to that of the homologous antigen". But then, he felt obliged to conclude that "the results of the partial saturation of precipitins with antigens ... give no conclusive evidence of the regular existence in a single immune serum of multiple antibodies which act specifically on various chemical groups of the antigenic proteins", as had been suggested by Ehrlich and Morgenroth. He went on to say that his results could be explained by a single crossreactive antibody for each antigenic protein. And further: "The peculiarities in specificity manifested by precipitinogens and agglutinogens suggest an essential difference in the chemical structures which determine the specificity of the two kinds of antigens." These conclusions were both unnecessary and wrong. They document a tendency among biologists to think that in nature only one or the other of two conceivable mechnisms can be right. Here, as in many other cases, both conceived mechanisms were correct, the crossreactivity of antibodies with similar target epitopes, and the existence in an antiserum of multiple antibodies to one protein.

The second example concerns faulty experimental work of two-times Nobel Laureate Linus Pauling. In 1900 James P. Atkinson and others rediscovered that the antitoxins are contained in the globulin fraction of a diphtheria antiserum[48], a finding first reported a decade earlier for tetanus antitoxin by G. Tizzoni and G. Cattani[14,15]. A number of subsequent investigators repeated and extended these experiments, but not much further progress was made because the salt fractionation methods in use were not able to do much better than to separate serum proteins into albumin and globulin. This only changed when Arne Tiselius in Uppsala invented the method of electrophoresis, which separated the serum proteins into four fractions, albumin, α-, β- and γ-globulins[49]. Elvin Kabat joined him, bringing a number of antisera along, and in 1939 they published a paper in which they reported that most antibodies are found in the γ-globulin fraction of these sera[50]. Salt fractionation procedures improved subsequently, mainly by the ammonium sulfate method of E.J. Cohn, by which antibodies were to be found among the γ-globulins as well, namely in Cohn's fraction II[51].

The multiple binding specificities of antibodies seemed hard to reconcile with the finding that they all belonged to a single type of molecule, γ-globulin, and so a number of theories were made to solve the dilemma. Even before this problem had come up, F. Breinl and F. Haurowitz (1930)[52] and S. Mudd (1932)[53] had postulated that the specific configuration of the antibody arose because the antigen acted as template around which the protein-synthesizing system made new antibody. Subsequently Linus Pauling (1940)[54] brought forward a slightly different theory, in which the formation of antibodies was regarded as due to refolding of preformed γ-globulin in the presence of the antigen, thus acquiring specificity. These and similar notions were later lumped together as the "instructional theories".

In 1942, Pauling and D.H. Campbell published a paper in which they presented experimental evidence seemingly proving Pauling's theory[55]. They used γ-globulins of normal serum purified according to Cohn's fractionation. The γ-globulin was denatured using various procedures including sodium hydroxide treatment or heating. The denatured γ-globulin was then incubated with various antigens and allowed to slowly renature in their presence. During renaturation precipitates appeared which where shown to contain both the antigen and some of the γ-globulin. Precipitates could be redissolved and the "manufactured" antibodies, after removal of the antigen, could be shown to specifically bind the same but not other antigens. There were only a few slightly disturbing findings, for example albumin could also be instructed to bind antigen, though to a lesser extent, and manufactured antibodies precipitated their antigen at somewhat different pH and temperature conditions than normal antibodies. Nevertheless, the authors concluded to "have succeeded in endowing normal serum globulin with the properties of a specific antibody; in other words, we would seem to have converted normal globulin into antibody".

While Campbell, without Pauling, reported in 1948 on his unsuccessful attempts to reproduce these experiments[56], Pauling was awarded the Nobel Prize in Chemistry in 1954, for his discoveries concerning protein structure. The instructional theories continued to prevail in the antibody paradigm, not only because nobody knew a better solution, but also because of supporting experimental evidence, including my third example. Rodney Porter, British biochemist and interested in antibodies, used Sanger's technique of amino-terminal amino acid sequencing, novel in 1950, to compare specific antibodies against ovalbumin with inert γ-globulin, both isolated from the same rabbit antiserum[57]. He envisaged two possible results: If the antibody differs from the inert γ-globulin, the specificity is likely determined by the primary structure; if they are the same, specificity is presumably conformational and Pauling is right. What he found was that both antibody and γ-globulin had the same amino terminal pentapeptide sequence. He reasoned: "As 19^5 pentapeptides could theoretically occupy the terminal position it is clear that the similarity between the biologically active and inert fractions cannot be coincidental. It therefore seems possible that this similarity will extend to a considerable part of the whole molecules...". Being a thorough and careful investigator, he continued: "...the combining sites which appear to be small may well have quite different composition from the equivalent section of the inert material". In spite of this caveat, Porter concluded: "The results described are in agreement with Pauling's theory of antibody formation in that no chemical distinction between the fractions could be found", and "it is clear that the chemical evidence described here ... is in accordance with Pauling's (instructional) theory".

In the 1950s at least 15 different theories were put forward to explain antibody formation and specificity[58], most of them now forgotten. What remained was the clonal selection theory of antibody formation of Macfarlane Burnet which, based on Jerne's natural selection theory, destroyed the instructional theories[59]. In 1959, Porter published his characterization on the enzymatic digestion fragments of γ-globulin[5], shortly thereafter Gerald Edelman reported the subunit structure of γ-globulin as consisting of heavy and light polypeptide chains[6,7]. These were perhaps the most pivotal contributions, after nearly seventy years of scientific fiction, which established antibodies as a scientific fact. Thereafter, only a short while – fifteen years – had to pass until Köhler's and Milstein's invention of the hybridoma technique established antibodies as a fact in the real world[60].

Chapter 17

The enticing network: Fiction forever

> *A scientific bandwagon is a situation in which large numbers of people, laboratories, and organizations rapidly commit their resources to one approach to a problem.*
> Joan H. Fujimura, in: A Pickering (ed) (1992)
> *Science in Practice and Culture.* Chicago

As discussed in chapter 7, in the early 1970s immunologists felt that a novel theory of adaptive immunity was badly needed. A first attempt was made in 1970 by Bretscher and Cohn, the "associative recognition" or "two-signal" hypothesis[1]. The citation rates indicate a growing attention until 1974 when Jerne launched his network theory[2]. Thereafter, Jerne's citation rates rapidly rose to impressive levels, with an anti-parallel drop of that of the two-signal hypothesis to near insignificance. Only after the decline of the network paradigm, as of 1990, citation rates of the two-signal hypothesis recovered to intermediate levels which are maintained until recently (chapter 9, Fig. 9.4).

In chapters 8 and 9 I have discussed a number of scientific reasons for the initial success of the network theory. Here I will concentrate on some social circumstances contributing to its attraction. The reader will find that the scientific and social origins of criteria for theory choice are not always clearly distinguishable, the boundaries not precisely delineable. Criteria of obvious social origin include the charisma of the teacher, the enticing feeling of the junior scientist to belong to a small esoteric circle and to be in possession of a deep-reaching insight, the promise of scientific success and of fame in its wake. Social criteria with a scientific touch include the expectancy to make a contribution to medicine, both from a charitable and from a monetary viewpoint. Properties of a theory with a strong appeal to unconscious human desires include a moderate degree of mysteriousness and some relatedness to ideas that appear to be "in the air", novel ways to view the world in general, beyond science.

There are few who did not sense Jerne's charismatic aura, both as a scientist and a human being. A detailed account can be found in Thomas Söderqvist's biography[3], *Science as autobiography; the troubled life of Niels K. Jerne*, which alludes in particular to Jerne as a man with a host of personal problems. While no attempt will me made here to elaborate further on Jerne's personality, I might add some remarks on the impact that an encounter with Jerne could have on a young scientist like myself. I met Jerne first during the early phases of the Basel Institute of Immunology, founded by Jerne and rapidly growing to the status of a temple to which one had to go for a pilgrimage at frequent intervals. In addition to the regular hiring of scientific staff, the Basel Institute had a policy to invite leading immunologists from all over the world to spend research sabbaticals of several months or even years. As a result, not only was there a unique collection of cutting edge experts present at all times, one also could meet and talk to one's most admired masters, heroes of science whose names one knew from reading the literature, but whom one so far had seen only at congresses from a distance as speakers behind the lectern, oneself being an anonymous part of a big audience in a huge lecture hall.

All of this created a unique aura of eminence in the institute, a center of superb intellectual standards which one entered, on the one hand, with the fear of exposing one's own insufficiency but, on the other hand, the pride of those privileged with access to a temple. Jerne himself was rarely seen in the laboratories, but occasionally did one get the opportunity to visit him in his office, the inner sanctum of the temple, in which he resided among long tables covered with row after row of manuscripts and reprints. The respect one felt was immense, was he not the creator of the selectional theory of immunity, the leading paradigm in the field*. Access to his office usually was granted because Jerne had somehow learned of an experiment that the visitor had performed, and which he found interesting. He would lead the conversation by asking about it, going into the smallest of detail such as what types of pipettes one had used, what strains of mice and how many per group, doses used for immunization, etc. When it came to one's interpretation, he did not seem very critical but mostly prepared to accept it, likely because Jerne selected his visitors because he liked their work in the first place. Conversely, the young visitor was left with a most positive impression, both of his own work as well as of Jerne as a teacher. He seemed able to absorb an incredible amount of information, and capable to distill the absorbed material into a coherent set of conclusions. He seemed a thorough and straightforward thinker, more synthetic than analytic, using a simple language, and was not at all given to the complicated reasoning and incom-

---

* Jerne and Burnet are commonly given equal credit for the clonal selection paradigm. On closer inspection, as outlined in chapter 6, Jerne's contribution, the selection of preformed antibody by antigen[4], did not convince the scientific community until Burnet added the aspect of clonality[5].

The enticing network: Fiction forever

prehensible argumentation which many important scientists use in discussions with junior colleagues. One left Jerne with a self-asserted feeling, one was worthy of the great man's attention, one didn't even get the idea that mistakes could be made under his eyes.

Two memorable conferences took place on the topic of idiotypic network under the leadership of Jerne and the Basel Institute's scientific staff. The first, in 1976, heralded the steep rise of the network theory. The second, in 1981, marked the beginning of a plateau era but featured, among the continued rejoicing of the enthusiasts, first subtle signs of doubt and disintegration. The 1976 meeting took place in a hotel with a three star restaurant, Ousteau de Beaumanière, near Les-Baux-de-Provence in the south of France, and had only fifteen participants. It had the ideal atmosphere to initiate the birth of an esoteric inner circle, a group of experts that share a thought style and develop it further by creative interaction. For a young scientist, the experience of participation was equal to being knighted and admitted to King Arthur's Round Table. Jerne himself fostered such sentiments by, for example, writing the first name of one of the young participants, Darcy Wilson, as "D'arcy". While Jerne was the undisputed intellectual leader, the participants included several senior scientists of some standing, including Benvenuto Pernis, Alfred Nissonoff, and Hilary Koprowski, whose personal acquaintance to make was a privilege in its own right. The format of the conference was a round table discussion, chaired by Jerne, with just a flip chart to draw some scheme if urgently needed (see Fig. 17.1). Between sessions, one would walk in the garden in small groups, continuing the discussion or making scientific small talk. Three star food and wine finished the day.

From the beginning it was an undisputed fact that the immune system was organized as an idiotypic network. Jerne in the introduction: "…these millions of different antibody molecules had been shown to be good antigens themselves, to constitute a universe of antigens present in the very elements of the immune system itself. The system thus has the makings of an autonomous network; the question is, to what extent would idiotypic regulation dominate its function". In other words, it was only the *extent* of dominance that remained to be determined. Throughout the discussion, experimental evidence in support of the key elements of the network was unanimously presented by all participants, including the coexistence of idiotype and anti-idiotype, the idiotypic relatedness of antibodies and T cell receptors, idiotype-specific helper and suppressor T cell regulation, and so on. A verbatim transcript of the Beaumanière meeting was unsuccessfully offered to a number of science publishers and finally published by the Basel Institute very much later, in 1980, under the title: "Idiotypes, what they said at the time"[6]. In the entire transcript one cannot find a single caveat or word of caution from any of the particians, and very little controversial discussion. In his foreword, written in hindsight, Jerne called it: "…a snapshot taken at an early stage of idiotype research" and: "The document shows

Figure 17.1
Niels Jerne listening to the author presenting data at the meeting at Les-Baux-de-Provence, April 1976. The proceedings of the meeting appeared in Schnurr I (Ed): *Idiotypes, what they said at the time. A discussion at Les-Baux-de-Provence, 2nd and 3rd April, 1976.* Basel, 1980.

that science is not a deductive exercise of moving from one logical conclusion to the next experiment, but that it is imagination in action, patching scraps of knowledge into something rich and strange". Obviously, Jerne knew what he was talking about.

The whole event was financed by the Mérieux Foundation, the charity fund of the major French pharma company engaged in vaccine development, a circumstance that gave rise to the comfortable feeling that something worth investing into was going on, and precluded putative ideas that the subject, idiotypic network, might perhaps just be an intellectual l'art pour l'art exercise. Idiotypic vaccination strategies for the treatment of infectious diseases or cancer were high flying ideas *in statu nascendi*, and the interest of a prominent industrial organization strengthened these hopes as realistic expectations. Success in the practical application of the network concept in medicine would secure both scientific fame and economic profit. The commercial exploitation of intellectual property, now common practice, was in its infancy at the time but the possibility certainly began to appear as an option to be considered. The participation in a small meeting at which these ideas were first born seemed to warrant a cutting edge position in the competition for priority likely to ensue. For example,

Hilary Koprowski, who had said very little at the meeting, soon thereafter began to develop monoclonal antibodies for idiotype-based vaccines for rabies and intestinal cancers[7,8]. Results of pre-clinical and clinical trials were not only published in scientific journals but also launched in the public press. The approach raised high expectations for a number of years, before it died away.

The second meeting on idiotypes took place in the Basel Institute[9]. It featured the format of a workshop with some two dozen invited speakers and more than a hundred participants in the packed audience. While the interest was overwhelming, the meeting lacked the esoteric character of the first one. Among the invited speakers were network enthusiast and opponents, as well as geneticists and structural biochemists interested in the 3-dimensional structure of antibody V regions and the genes encoding them, most of them with just marginal interest in the network. Scientific advances were reported mostly by the latter group, whereas network research seemed just more of the same, additional data but no novel perspectives. There are subtle indications in the phrasings of some leading network researchers that their convictions had begun to be shaken. For example Jerne, who once again introduced the network theory by going over the Ab1-Ab2-Ab3-terminology, remarked: "If you tell me during the meeting that there is some basic mistake in all this, I will be surprised". Jacques Urbain, another prominent network researcher, began his talk like this: "I would like to discuss three points: Is the network a logical necessity? Is the network an empirical necessity? And, is the network a necessity of evolution? With regard to the first question, the answer is most probably yes. For the second, the answer is also yes, but it is still necessary to leave an interrogation mark. As to the third question, my answer is yes, but I think that some – if not all – of you will disagree". Melvin Cohn, always the most articulate network opponent, predictably made his point: "The immune system is not interested in the idiotypic specificity. The immune system is interested in the antigen combining site, not in the idiotypic specificity of the receptor", and "...the immune system must eventually deal with a pathogen". The enticing network had begun to lose some of its fascination. The opposition, which was feeble at first, had gained in strength, and had succeeded in planting a seed of doubt in the proponents' minds. Perhaps most importantly, molecular biology and structural biochemistry began to yield stunning new information that didn't need an idiotypic network, it rather seemed to put it aside as unnecessary ballast on the way to a new set of paradigms (see chapter 13).

Not only were the social circumstances around the beginnings of the network theory enticing, also the theory itself had certain assets that appealed to the non-scientific side of the human mind. One of these was a somewhat mystical touch, resulting from the fact that certain essential elements of the theory escaped a clear scientific delineation and therefore remained in the realms of mysticism. Many scientist colleagues will vehemently reject the idea that a mystical quality could contribute to the attraction of a scientific notion, but is it not true that the human mind has been fascinated by mysticism at all times? Scientists might claim to be less given to the fascination of mysticism than the ordinary human, but no one will convince me that they – as a group – are fully resistant.

Mystical elements of the network theory included the exceedingly large numbers of idiotypes present in the system, commonly expressed in exponential numbers because decimal figures would require impractically long rows of zeros. Beyond everybody's imagination, seemingly endless, only to be roughly estimated but never exactly enumerable, the repertoire of idiotypes in the system was proposed to be "complete" by Coutinho[10]. Completeness is akin to infinity, a term used to describe entities like the universe or the wisdom of God, subjects of awe or faith. Related to the notion of completeness was the notion of "internal image", meaning that the idiotypic repertoire encompassed images as well as mirror images of virtually all structural elements in the environment, existing and to be existing. The entire molecular world was repeated within the body itself. Again a notion beyond anyone's realistic comprehension, and thus all the more fascinating. The adjective "complete" and the noun "internal image" were among the most frequently used terms in network literature.

One can argue that the notion of infinity may be applied to the antibody repertoire as such, independent of whether or not one views it as an idiotypic network. Indeed, immunologists have frequently used the acronym GOD as short for "generator of diversity" in pre-network times, an intended ambiguity in reference to the then mysterious genetics of antibody diversification. While some may insist that this was merely a joke, to me it means that fascination with mysticism is not unique to the network theory but a common, though non-admitted, driving force in science. The abundant use of the phrases "complete repertoire" and "internal image" in the context of the network theory documents just a particularly obvious example.

An intriguing asset likely having contributed to the success of the network hypothesis was the term "network" itself. Few other terms have enjoyed a similar success in the spoken and written language in recent years, as documented by the increasing frequency of hits upon searching the major publication databanks covering all natural sciences, social sciences, and arts and humanities (Fig. 17.2). Of the well over 500,000 publications that contain the term "network" in the title or abstract only a few appeared between 1970 and 1980, with a shallow linear increase in frequency. Thereafter an explosion took place with an exponential increase which continues until today. The term is used in multiple contexts comprising all types of living communities, politics, economics, ecology, biological systems, etc. While the meaning varies somewhat according to context, it nearly always applies to complex interacting multispecies systems of many different types. In chapter 8 I have discussed the role of cyber-

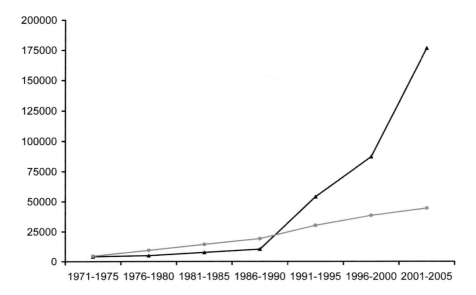

Figure 17.2
Numbers of publications/5 year period containing the term "network" and "competition" in the title or abstract.
The graph shows the dynamics of publications referring to "network" (black triangles) compared to that referring to "competition" (grey circles) between 1971 and 2005. Both terms are used in multiple contexts in many different scientific disciplines, resulting in huge numbers of hits for both terms. The shallow and steady increase in publications referring to competition roughly reflect the overall increase in scientific publications during the period analyzed. In contrast, publications referring to "network" show a sudden and steep increase after 1990 which continues until today, reflecting the growing general interest which networks receive in many scientific fields (Data extracted from Thompson Scientific Web of Science).

netics as a proto-idea for the idiotypic network theory, notably second order cybernetics in which self-regulating networks are an essential element. The increasing popularity of the term suggests that the notion of network seems to satisfy a common need in efforts to understand the complex structures of the environment in which we live, in a very general sense. The immune system is such a complex structure and it is no surprise that a network concept of its function was similarly popular as in other areas of human intellectual activity. Indeed, with respect to timing, the idiotypic network theory appears as a forerunner, even anticipating the success that network concepts of complex structures were going to enjoy in the future.

Not only were social phenomena involved in the initial success of the idiotypic network theory, they also had a role in its subsequent rejection. The most drastic case is the tabuization of the suppressor T cell in the community of immunologists, heralding the post-network era*. While there is no doubt that in the former research on suppressor T cells mistakes have been made (chapter 11), the recent overwhelming success of suppressor cells under the new name of "regulatory T cell" can only mean that the former notions on T cell-mediated suppression were not entirely erroneous. They contained correct elements, decorated with phantasies of the investigators, the latter perhaps more pronounced than usual but in principle not different from other scientific notions. Nevertheless, a consensus developed in the scientific community to exclude the subject from further consideration, as a topic of conference lectures, publications, grant applications, not to speak of research projects in the laboratory. The situation was the exact reverse of a scientific bandwagon, in this case one deserted of passengers after the crash. Scientists participating in the few exceptional activities had to excuse themselves for doing so, such as Douglas Green and David Webb, who in 1993 wrote a report on an international (!) workshop with 55 (!) participants, on immune suppression[12]. It begins: "There is little doubt that the S-word (suppression, as in suppressor T cells) is the nearest thing to a dirty word we have in cellular immunology. Its use is considered by some (not all) to be synonymous with overinterpretation of scanty data and phenomenology bordering on the mystical.... Despite the fact that many mechanisms of T cell-mediated suppression are not controversial, ... there continues to be a stigma associated with research in this area". There is no doubt that the ban of suppressor cell research in the 1990s was a social phenomenon that, while originating from scientific disappointment, as a reaction far exceeded any reasonable proportion.

While the ban of suppressor cell research had an initiating role in the decline of the entire network paradigm, rejection of the latter was less radical, more like a slow fading away, in the wake of a growing realization among immunologists that further advances in the understanding of immunity were not to be expected from the network. As a result, the network paradigm gradually ceased to direct mainstream immunological research but it did not die a sudden and complete death, it survived even until today in certain quarters such as clinical research in which, for example, the notion of internal image continues to govern cancer vaccination attempts (see chapter 13).

The scientific developments leading to the eventual rejection of the idiotypic network paradigm have been discussed in detail in Part II, chapters 10–12. It would be too simple to conclude, however, that the rejection of the network paradigm was entirely due to the erroneous notions derived from certain areas of network research such as the T cell receptor or suppressor

---

* A critical account of this process has been given by Keating and Cambrosio[11], who analyzed the difference in acceptance of helper and suppressor T cells from the viewpoint of the occurrence of scientific artifacts. With the new information on Treg, one cannot maintain that suppressor T cells have been artifacts. They have been a normal fictional scientific notion, though with an exceptionally high proportion of fiction.

T cells. Against this assumption stands the fact that much of research under the network paradigm was correct, robust, reproducible in many laboratories. Successful scientific endeavors such as the development of the antibody paradigm (chapter 16) were accompanied by just as many errors and misconceptions, but nevertheless yielded solid knowledge that today serves as reliable guideline for multiple human activities. Some colleagues like to think that with the advance of scientific technology the frequency of errors decreases, resulting in an increasing probability of scientific notions turning into fact[13]. This can also not explain the difference between the network and antibody notions, as the latter was studied at a much earlier time, with analytical methods much less developed than in the era of network research. What then were the critical factors that distinguished the network from the antibody paradigm. Or, more general, what distinguishes scientific notions that turn into fact from those that don't. Answers to these questions will be sought in the following chapter.

Chapter 18

Logic and laws in life science

Pure logic is not the only rule for our judgements; certain opinions which do not fall under the hammer of the principle of contradiction are in any case perfectly unreasonable.
Pierre Duhem (1987) *Prémices philosophiques.*
Brill Academic Publ.

From David Hume[1] to Karl Popper[2], science theoreticians have struggled with the problem of induction: Scientists usually are convinced that conclusions deduced from scientific methodology, if they have proven to be correct over and over in the past, will continue to hold true in the future. Philosophers, in contrast, insist that according to the principles of formal logic one can never be sure about the validity of an empirically deduced scientific fact at a later time or another place. The best one can say is that among the many possible theories that can be deduced from empirical evidence, the one that has proven the most successful in the past is likely to continue being successful in the future[3]. As a way out of the dilemma, Popper has suggested that scientists should refrain from declaring their theories, even the most successful ones, absolute facts. They should rather accept them as long as they are successful, and be prepared to drop them if they fail to deal successfully with upcoming recalcitrant evidence[4]. As discussed in Part I, subsequent theoreticians have realized that science does not work that way, and so the lack of a logical foundation of inductive knowledge remains a dilemma.

Some philosophers have defined certain exceptional categories of science which do not fall under the doubt of induction, notions that do not have to be derived empirically but are evident in themselves, so-called axioms or postulates. For example, Hume accepted Euclidian geometry as *a priori* true, based on a system of self-evident axioms that cannot be challenged by logical reasoning[5]. Kant included Newtonian mechanics into this category[6,7], but we know today that Kant was wrong. Newton's mechanical

Logic and laws in life science

laws are sufficiently accurate only for relatively small objects and their motions, when the dimensions of space on earth are taken as the absolute category. Newton's laws fail in outer space, were according to Einstein's theory of relativity the velocity of light is the only absolute category and everything else including space is relative. Some physicists maintain that Newton's mechanics remain correct for motions on earth, on which for all practical purposes one can neglect Einstein's relativity theory. With equal justification, others insist that Newton was inaccurate, though to a marginal extent as far as motions on earth are concerned. Euclidian geometry has been questioned as well. If we follow Joan Fujimura[8], observations of the German mathematician Friedrich Gauss in the 19th century challenged Euclid's postulates*. When working as a geographic surveyor in Bavaria Gauss observed that Euclid's two basic axioms were invalid on the curved surface of the earth, i.e. the angles of a triangle did not add up to 180°, and the ratio of the circumference over the diameter of a circle deviated from the constant, π. Whether or not Gauss's observations indeed invalidate Euclidian geometry seems at least doubtful, and Gauss himself seems not to have meant it that way. The reason for giving these examples here is to illustrate that arguments can be found that query the validity of even very fundamental axioms in mathematics and physics. When examined under a different set of conditions, facts may cease to be facts. If this is the case even for the most quantitative and exact sciences, what about biology, a predominantly qualitative science in which accurate measurements are the exception rather than the rule?

Mathematics is a science based on abstract laws which are unconditionally valid, independent of time and place. Physics, using mathematics as a major tool, aims at defining equations according to which the material part of nature functions. Not only physicists refer to these equations as laws, meaning that they are without exception and unconditionally valid. The above example shows that Newton's laws of mechanics, historically even taken as axioms, have been degraded by Einstein's theory of relativity from laws to rules, valid under a set of defined conditions but not in general. Other examples can certainly be found, but it is not for me to discuss physics, of which I hold great respect. I am concerned with the corresponding aims of biology. Can the scientific facts that develop from biological notions be compared to the laws or rules of physics, or do they fall behind with respect to general validity? In the latter case, how far do they fall behind? Or is general validity perhaps a category that is not adequate in biology?

In chapter 3 I have discussed several examples showing that claims of general validity in biology rarely survive for long, they very often run up

---

* Fujimura used this example in an article[8] on Science Wars (see chapter 5), in which she argues that Gauss's heretical "non-Euclidian geometry" was suppressed at the time, just as Alan Sokal's hoax, with the ridicule it generated, was an attempt to censor science critiques.

against observations of exceptions. Related to this is the experience that scientific controversies, of which there have been many particularly in immunology, often have been resolved by the realization that both sides have been correct, if only partially. What had appeared as mutually exclusive alternatives turned out to be coexisting solutions of a problem in nature. This is not to say that all of these alternatives are encountered with equal frequency, mostly one is dominant and therefore often taken as the only existing solution. A recent example is the detection of a rare alternative version of generating variable lymphocyte receptors in jawless fish, an early class of vertebrate of which only two orders are known today, lamprey and hagfish. Lymphoid cells of these animals possess a pair of related genes into which diverse variable leucine-rich repeats are inserted, thus generating a receptor repertoire of pronounced diversity[9]. The mechanisms for the generation of diversity of these genes, as well as the genes themselves, are entirely different from the gene segments in higher vertebrates which are rearranged to generate B and T cell receptors. Yet both systems result in diverse receptor repertoires for antigen recognition. As indicated by their limited numbers of contemporary orders and species, jawless vertebrates have had limited success in evolution, and one can speculate that this was partially due to a lower efficiency of their version of variable lymphocyte receptors, compared to that of antibodies and T cell receptors of jawed vertebrates. Nevertheless, jawless fish survived until today suggesting that their version of immunity was efficient to some extent. Importantly, the notion that the adaptive immune system evolved by generating antibodies and T cell receptors for antigen recognition, seemingly a law until recently, is now merely a rule to which there are exceptions.

The examples discussed above (and in chapter 3) force one to accept that the establishment of a fact in biology cannot be used to formulate a law, if a law is defined as a statement which is valid without exception, once and forever. As pointed out repeatedly in this text, this does not mean that biological facts are not facts. So the question arises whether or not general validity is a useful criterion to assess the value of biological facts. Once a fact in biology has passed its trials of strength over time, it is included into the knowledge of mankind and can be used to manipulate biology to human advantage. Further examples for the innumerable instances in which this has happened need not be listed here. To a biological fact belongs the knowledge about where it applies and where not, though this information is often added with delay, once the first exceptions have become apparent. Most experimental biologists would agree to the notion that there is no need for a biological fact to apply in all possible instances. One of the most fundamental principles in biology is the universality of the genetic code, i.e. the relationship of the nucleotide triplets of DNA to the amino acids in polypeptide chains, which is the same in all living organisms on earth[10]. Some triplets can vary in the third nucleotide, but one of these variants is always preferred compared to the others[11]. Recently, the multiple attempts to express mammalian genes in bacteria have taught us that the preferences among such alternative triplets are not universal, they differ between mammals and bacteria[12]. For efficient bacterial expression of mammalian genes, investigators now exchange those triplets for the ones preferred in bacteria. The example shows that even the most general principles in biology run up against modifications, if only one looks hard enough. But this does not invalidate them. On the contrary, knowledge expands.

The experience that biological facts are not universal forces one to conclude that multiple solutions exist for each of the problems that living organisms face in their environment. Evolution has tested a large variety of available solutions and those that we encounter by studying model organisms have been selected because they work. Available in the evolutionary context means that a solution offers itself for testing because a number of genes have been generated whose protein products, working in concert with existing proteins, offer a solution for one of these problems. The large variety of organisms living within a habitat clearly documents that there are many feasible solutions to a problem, and this not only applies to entire organisms but to their bodily functions as well, immunity against infection being only one of them. In evolutionary testing only one result counts: efficient reproduction. If one way of solving a problem is dominant in nature, this only means that it has proven effective in supporting the multiplication of the organism using it, and more so than all the putative alternatives that evolution may have tested in the past. But more often than not, more than one solution can solve a problem, if not equally well, as observed for the case of variable lymphocyte receptors described above.

Novel genes and proteins to be tested in evolution are the result of random mutations in existing genomes and, before their existence, their nature and properties are unpredictable. They are not designed to fulfill a particular purpose, they are randomly generated and then tested for whether or not they do. It is not at all a matter of course that the dominant solutions are the absolutely optimal ones, they are just the best of those that have become available for selection. The number of possible mutations is seemingly endless, as is the number of environmental factors influencing their selection. Evolution is thus a mindless process, based on very large numbers of possible solutions tested during very long periods of time, but without a defined goal. Evolution may seem to be directed at the generation of higher organisms, such as *Homo sapiens*, but this directionality is merely a consequence of the test parameter, efficient reproduction, which stabilizes viable solutions. The way evolution comes to solutions is thus very different from the way the human mind tries to solve problems, namely by goal-oriented thinking. Solutions for biological problems, designed by the human mind, almost never coincide with that selected by evolution. Since theories in biology are nothing else but solutions of biological problems designed by the human mind, they rarely correctly anticipate what is later found by experiments. Theodosius Dobzhansky, the famous evolutionary biologist, has written a lucid article on this dilemma[13], entitled: "Nothing in biology makes sense except in the light of evolution".

Dobzhansky's title may be rephrased to read: "There is no logic in biology except for the logic of evolution". The meaning of "logic" is a subject of study in philosophy all by itself, and no attempt is made here to add to this subject. Many philosophers, including some of the most influential, have elaborated on what logic really means, with multiple definitions and changes in meaning over time. Here I am using it simply as synonymous for straightforward reasoning, as in the Aristotelian example: If two correct sentences, such as: "all men are mortal" and: "all Greeks are men", share a term (men), a third correct sentence follows: "all Greeks are mortal".

Scientists rely on logic when they deduce theories from experimental observations, but in biology these deductions very often fail. For example, to explain the observation that, within one class of antibody polypeptide chains, all the millions of different variable regions are associated with the same constant region, the "two genes – one polypeptide" hypothesis was developed[14]. For the case of antibodies this hypothesis seemed a logical modification of the previous notion, valid for all other proteins: One gene – one polypeptide. When the techniques became available to investigate the nature of antibody genes, the solution which was found was entirely different. While the hypothesis was correct as far as the constant region genes were concerned, nobody had – or could have – predicted that antibody V genes are recombined out of two or three segments, which was the solution that evolution had selected[15]. Indeed, very few biological mechanisms have been correctly predicted by theories. Who would have predicted that eukaryotic genes have introns, or that the human genome consists of less than 30,000 genes, so that epigenetic mechanisms have to be summoned in order to account for phenotypic diversity? After such stunning observations have been made, scientists often engage in reconstructing the ways in which evolution has selected them. Indeed, the selective advantages of a discovered mechanism often seem strikingly obvious in retrospect. Although based on conjecture, one can mostly intuitively understand why a solution has been selected. For example, an advantage of having introns in eukaryotic genes is that recombination, an important asset of sexual reproduction, does not destroy a gene when it takes place in an intron, even if the recombination has been inaccurate. Introns thus make the genome both more flexible for generating new genes and more resistant to gene loss by sloppy recombinations. Nevertheless, nobody would have predicted introns in eukaryotic genes before they have been encountered by experimentation. Evolutionary solutions have rarely been correctly anticipated.

Whatever is established as a fact in biology cannot be derived by logical reasoning, it can only be retrospectively understood as logical in evolutionary terms, if evolution can be accepted as a form of logic. The uncertain role of logic in biology is reminiscent of a discussion that arose between physicists and philosophers, and among physicists themselves, in the context of the impact that quantum mechanics had on views of the world in general. The theory of quantum mechanics has a highly disturbing consequence, namely the uncertainty relationship, which means that of the two properties of an elementary particle, its location and its impulse, only one can be determined at any one time, the other remains uncertain. Werner Heisenberg reasoned that the sentence: "If we have exact knowledge of the present, we can predict the future", can no longer be upheld because the uncertainty relationship teaches us that it is impossible to have exact and full knowledge of the present. He concluded that therefore the "principle of causality is definitively invalidated"[16]. Carl Friedrich von Weizsäcker went even further in suggesting: "It would be conceivable that the example of present day physics obviates structures of existence which would be incompatible with the ontologic hypothesis, which is the basis of classical logic", and considered the necessity of a new form of logic, so-called quantum logic, which would include classical logic as a special case[17]. Needless to add, these statements have been vehemently rejected by philosophers who, understandably, insist on the validity of the principles of causality and of formal logic, the latter according to Gottfried Wilhelm Leibnitz being "valid in all possible worlds"[18].

It is possible that the reason why logic fails in biology is not a matter of principle but a consequence of our intellectual inability to handle problems with more than a limited number of independent variables. If we could construct algorithms that calculate all possible genetic and epigenetic modifications of, say, a simple virus, and evaluate all of them against all constant and variable parameters in its habitat, and if we could construct the computers that could handle such algorithms, predictions of the evolutionary future of that virus could perhaps be possible. Indeed, attempts in this direction are being undertaken[19,20]. For example, computer programs have been developed for predicting the genetic diversification and the immunological selection of certain variants of HIV in infected individuals[21]; or scientists have developed evolutionary algorithms which companies use for drug design[22], and so on. These approaches have yet to stand the trial of strength in the future. Whether or not such computer-aided attempts to predict evolutionary solutions will be successful in the future, the answer will not affect the conclusion that formal or classical logic is inappropriate for biology. The logic of biology is the logic of evolution.

The inability to predict solutions of biological problems by pure logic does not mean that theories have no room in the life sciences. Theories in biology may be divided into two not sharply distinct categories, predominantly deductive and predominantly inductive ones. A deductive theory aims at connecting a novel set of experimental observations with current paradigms. More often than not, novel evidence gives rise to notions that appear to be at variance with, or unconnected to, any of the theories in a given paradigm. Theories are thus adjusted to the new evidence, often without destroying the existing paradigm but sometimes drastic enough to cause a paradigm shift à la Kuhn. Important is that a deductive theory starts with an experimental observation, it is triggered by a piece of evidence that is so far not readily connected to the current paradigm. A deductive theory is made after such evidence has appeared and deals with it in retrospect. Because many deductive theories are restricted to small sectors of biology, using the term theory may even be debatable in some cases. In contrast, what might be called an inductive theory usually covers a comprehensive sector of biology. Moreover, inductive theories are often put forward independently of novel evidence, they represent a novel concept of rationalizing the existing evidence, what is novel about them is the reasoning, not the evidence. As such, an inductive theory is in danger of violating Occam's razor[23], as there may be no compelling need for its development. An important property of an inductive theory is its forward element. By proposing a new way of connecting existing notions, it triggers experiments that have not been done before and predicts their results. In that respect, inductive theories can be very productive and stimulating elements in the progress of a scientific field.

Deductive theories are relatively safe, and immunology is full of successful examples. In contrast, inductive theories are risky. The fifteen or more theories on antibody formation put forward in the late 1950s (see chapter 16) were inductive theories, trying to make sense of an inconclusive body of existing evidence, without novel informative evidence that could have pointed the way. Among those, only Burnet's clonal selection theory survived, predicting the clonal expression of antibody genes for which there was no direct evidence at the time, but which appeared to make sense all things considered. Inductive theories are risky because of the evolutionary conditioning of logic in biology. Nevertheless, Burnet's example shows that inductive theories, too, can prove to be correct, though they rarely are. I know of no other inductive theory in immunology that was even vaguely correct, from Ehrlich's side chain theory to the various versions of predicting the genetic basis of antibody diversity, only to name a few. Indeed, it seems that biologists have given up on inductive theories, certainly in immunology none has been formulated since more than 30 years. The last inductive theory in immunology was the idiotypic network theory of the immune system. It had all the pitfalls of its kind, it tried to predict evolutionary solutions, using logical reasoning, and it violated Occam's razor by making more than a minimum of assumptions. What a pity it failed!

References and further reading, Part III

Chapter 15

References

1 Popper KR (1959) *The Logic of Scientific Discovery.* Hutchinson, London
2 Kuhn TS, (1962; 1970 2nd edition, enlarged) *The Structure of Scientific Revolutions.* University of Chicago Press
3 Fleck L (1979) *Genesis and Development of a Scientific Fact.* Ed. by TJ Trenn and RK Merton. The University of Chicago Press, Chicago/London
4 Medawar PB (1969) *Induction and intuition in scientific thought.* American Philosophical Society, Philadelphia
5 Silverstein AM (1989) *A history of immunology.* Academic Press, Inc. San Diego, New York
6 Laudan L (1996) *Beyond Positivism and Relativism. Theory, Method, and Evidence.* Westview Press, Oxford, UK
7 Knorr Cetina K (1981) *The Manufacture of Knowledge. An Essay on the Constructivist and Contextual Nature of Science.* Pergamon Press, Oxford
8 Rheinberger HJ (1992) Experiment, difference, and writing: I. Tracing protein synthesis. *Stud Hist Phil Sci* 23: 305

Chapter 16

References

1 Avery OT, MacLeod C, McCarty M (1944) Studies on the chemical nature of the substance inducing transformation of pneumococcal types. *J Exp Med* 79: 137–158
2 Watson JD, Crick FH (1953) Molecular structure of nucleic acids; a structure for deoxyribose nucleic acid. *Nature* 171: 737–738
3 Nirenberg MW, Matthaei JH (1961) The dependence of cell-free protein synthesis in *E. coli* upon naturally occurring or synthetic polyribonucleotides. *Proc Natl Acad Sci USA* 47: 1588–1602
4 Silverstein AM (1989) *A history of immunology.* Academic Press, Inc. San Diego, New York
5 Porter RR (1959) The hydrolysis of rabbit γ-globulin and antibodies with crystalline papain. *Biochem J* 73: 119–126
6 Poulik MD, Edelman GM (1961) Comparison of reduced alkylated derivatives of some myeloma globulins and Bence-Jones proteins. *Nature* 191: 1274–1276
7 Edelman GM, Poulik MD (1961) Studies on structural units of the gamma-globulins. *J Exp Med* 113: 861–884
8 Lindenmann J (1984) Origins of the terms 'antibody' and 'antigen'. *Scand J Immunol* 19: 281
9 Behring E (1890) Untersuchungen über das Zustandekommen der Diphterieimmunität bei Thieren. *Dtsch Med Wochenschr* 16: 1145
10 Behring E, Kitasato S (1890) Über das Zustandekommen der Diphterieimmunität und der Tetanus-Immunität bei Thieren. *Dtsch Med Wochenschrift* 16: 1113

11 Behring E, Wernicke E (1892) Über Immunisierung und Heilung von Versuchsthieren bei der Diphterie. *Z Gesamte Hyg* 12: 10

12 Behring E (1894) *Das neue Diphteriemittel.* Berlin

13 Zeiss H, Biehling R (1941) *Behring, Gestalt und Werk.* Bruno Schultz Verlag, Berlin-Grunewald

14 Tizzoni G, Cattani G (1891) Über die Eigenschaften das Tetanus-Antitoxins. *Zentralbl Bakteriol Mikrobiol Hyg (A)* 9: 685

15 Tizzoni G, Cattani G (1891) Fernere Intersuchungen über Tetanus-Antitoxin. *Zentralbl Bakteriol Mikrobiol Hyg (A)* 10: 33

16 Gedoelst ML (1892) *Traité de microbiologie appliquée a la médecine vétérinaire.* J. Van In and Co., Lierre, Belgium

17 Mazumdar PM (1974) The antigen-antibody reaction and the physics and chemistry of life. *Bull Hist Med* 48: 1

18 Verworn M (1894) *Allgemeine Physiologie: Ein Grundriss der Lehre vom Leben.* Fischer, Jena

19 Graham T (1861) Liquid diffusion applied to analysis. *Phil Transl* 151: 183

20 Ehrlich P (1891) Experimentelle Untersuchungen über Immunität. II. Über Abrin. *Dtsch Med Wochenschr* 17: 1218

21 Wells HG (1929) *The chemical aspects of immunity.* Chemical Catalog, New York

22 Bäumler E (1984) *Paul Ehrlich, Scientist for Life.* Holmes and Meier, New York and London

23 Marquardt M (1951) *Paul Ehrlich.* Springer Verlag, Berlin

24 Ehrlich P (1987) Die Wertbemessung des Diphterie-Heilserums, und deren theoretische Grundlagen. *Klin Jahrb* 6: 299

25 Ehrlich P, Morgenroth J (1901) Über Hämolysine: fünfte Mitteilung. *Berl Klin Wochenschr* 38: 251

26 Ehrlich P (1900) On immunity with special reference to cell life. Croonian Lecture. *Proc Roy Soc London* 66: 424

27 Lie JT (1980) Weigert, Carl (1845–1904), a pathfinder in medicine. *Mayo Clinic Proceedings* 55: 716–720

28 Bordet J (1900) Les sérums hémolytique, les anticoprs et les théories des sérums cytolytique. *Ann de l'Inst Pasteur* 14: 257

29 Danyz J (1902) Mélanges des toxines avec des antitoxines. *Ann de l'Inst Pasteur* 16: 331

30 Eisenberg P, Volk R (1902) Untersuchungen über Agglutination. *Ztschr f Hyg* 40: 155

31 Schröder-Gudehus B (1978) *Les scientifiques et la paix.* Presses Univ Montrael, Montreal

32 Buchner H (1987) Die Bedeutung der aktiven löslichen Zellprodukte. *Münch Med Wochenschr* 44: 299–302

33 Van Dungern E (1903) *Die Antikörper. Resultate früherer Untersuchungen und neue Versuche.* Gustav Fischer Verlag, Jena

34 Ehrlich P, Morgenstern J (1901) Über Hämolysine, sechste Mitteilung. *Berl Klin Wochenschr* 38: 569

35 Cambrosio A, Jacobi D, Keating P (1993) Ehrlich's 'beautiful pictures' and the controversial beginnings of immunological imagery. *Isis* 84: 662

36 Le Dantec F (1916) Le bluff de la science allemande. In: Petit G, Leudet M (ed): *Les Allemandes et la Science.* Alcan, Paris

Références and further reading, Part III 243

37 Dean HR (1917) Mechanism of serum reactions. *Lancet* 13: 45
38 Deutsch L (1899) Contribution a l'étude de l'origine des anticorps typhiques. *Ann Inst Pasteur* 13: 689
39 Buchner H (1893) Über Bacteriengifte und Gegengifte. *Münch Med Wochenschr* 40: 449
40 Emmrich R, Loew O (1991) Über biochemischen Antagonismus. *Zentralbl Bakteriol Mikrobiol Hyg (A)* 30: 552
41 Pick EP (1912) Biochemie der Antigene, mit besonderer Berücksichtigung der chemischen Grundlagen der Antigenspezifität. In: Kolle W, Wassermann A (eds): *Handbuch der pathogenen Mikroorganismen.* Fischer, Jena, 685
42 Abel JJ, Ford WW (1907) On the poison of Amanita phalloides. *J Biol Chem* 2: 273
Ford WW (1908) Pathology of *Amanita phalloides* intoxication. *J Inf Dis* 5: 116
43 Heidelberger M, Avery OT (1923) The soluble specific substance of *Pneumococcus*, 2nd paper. *J Exp Med* 40: 301
44 Landsteiner K (1920) Specific serum reactions with simply composed substances of familiar constitution (inorganic acids). XIV. Announcement on antigens and serological specifity. *Biochemische Zeitschr* 104: 280
Landsteiner K (1922) On the formation of heterogenetic antigen by combination of hapten and protein. *Proceedings of the Kononklije Akademie van Wetenschapen te Amsterdam* 24 (1/7): 237
45 Ehrlich P, Morgenroth J: *Berl Klin Wochenschr* 1899: No. 1, No. 22; 1900: No. 21, No. 31; 1901: No. 10, No. 21
46 Landsteiner K (reprinted 2001; original 1901) Agglutination phenomena of normal human blood. *Wien Klin Wochenschr* 113: 768
47 Landsteiner K, Van der Scheer J (1924) On the specificity of agglutinins and precipitins. *J Exp Med* 40: 91
48 Hiss PH Jr, Atkinson JP (1900) Serum globulin and diphteria antitoxin. – A comparative study of the amount of globulin in normal and antitoxic sera, and the relation of the globulins to the antitoxic bodies. *J Exp Med* 5: 47–66
Atkinson JP (1900) The fractional precipitation of the globulin and albumin of the normal horse's serum and diphteria antitoxic serum, and the antitoxic strength of the precipitates. *J Exp Med* 5: 67–76
49 Tiselius A (1937) Electrophoresis of purified antibody preparations. *J Exp Med* 65: 641–646
50 Tiselius A, Kabat EA (1939) An electrophoretic study of immune sera and purified antibody preparations. *J Exp Med* 69: 119–131
51 Cohn EJ, McMeekin TL, Oncley JL, Neweu JM, Hughes WL (1940) *J Am Chem Soc* 62: 3386;
Cohn EJ, Luetscher JA Jr, Oncley JL, Armstrong SH Jr, Davis BD (1940) *J Am Chem Soc* 62: 3396
52 Breinl F, Haurowitz F (1930) Chemische Untersuchungen des Präzipitates aus Hämoglobin und Anti-Hämoglobin-Serum und Bemerkungen über die Natur der Antikörper. *Hoppe-Seyl Z Phys Chem* 12: 45
53 Mudd S (1932) A hypothetical model of antibody formation. *J Immunol* 23: 423
54 Pauling L (1940) A theory of the structure and process of the formation of antibodies. *J Amer Chem Soc* 62: 2643
55 Pauling L, Campbell DH (1942) The manufacture of antibodies *in vitro. J Exp Med* 76: 211

56   Campbell DH (1948) The nature of antibodies. *Ann Rev Microbiol* 2: 269
57   Porter RR (1950) A chemical study of rabbit antiovalbumin. *Biochem J* 46: 31
58   Campbell DH (1957) *Blood* 12: 589
      Coons HH (1958) *J Cellular Comp Physiol* 52 Suppl 1: 55
      Dubert JM (1959) *Ann Inst Pasteur* 97: 679
      Fishman M (1959) Nature 183: 1200
      Karush F, (1959) In: Shaffer J, LoGrippo G, Chase MW (eds): *Mechanisms of Hypersensitivity*. Little, Brown & Co., Boston, Mass.
      Koshland ME (1957) *J Immunol* 79: 162
      Koshland ME, Englberger F (1957) *J Immunol* 79: 172
      Najjar VA, Fisher J (1955) *Science* 122: 1272
      Pappenheimer AM Jr, Scharff M, Uhr, JW (1959) In: Shaffer JH, LoGrippo GA, Chase MW (eds): *Mechanisms of Hypersensitivity*. Little, Brown & Co., Boston, Mass, 417
      Schweet RS, Owen RD (1957) *J Cellular Comp Physiol* Suppl 1: 199
      Speirs RS (1958) *Nature* 181: 681
      Stavitsky AB (1957) *Federation Proc* 16: 652
      Lederberg J (1959) *Science* 129: 1649
      Boyden SV (1960) *Nature* 155: 724
      Monod J (1959) In: Lawrence HS (ed): *Cellular and Humoral Aspects of the Hypersensitive States*. Hoeber-Harper, New York, 628
59   Burnet FM (1957) A modification of Jerne's theory of antibody production using the concept of clonal selection. *Aust J Sci* 20: 67
      Burnet FM (1959) *The clonal selection theory of aquired immunity*. Cambridge University Press, Cambridge
60   Köhler G, Milstein C (1975) Continuous cultures of fused cells secreting antibody of predefined specificity. *Nature* 256: 495–497

Chapter 17

References

1   Bretscher P, Cohn M (1970) A theory of self-nonself discrimination. *Science* 169: 1042
2   Jerne NK (1974) Towards a network theory of the immune system. *Ann Immunol (Inst. Pasteur)* 125 C: 373
3   Söderquist T (2003) *Science as autobiography, the troubled life of Niels Jerne*. Yale University Press, New Haven, London
4   Jerne NK (1955) The natural selection theory of antibody formation. *Proc Nat Acad Sci USA* 41: 849
5   Burnet FM (1957) A modification of Jerne's theory of antibody production using the concept of clonal selection. *Aust J Sci* 20: 67
      Burnet FM (1959) *The clonal selection theory of aquired immunity*. Cambridge University Press, Cambridge
6   Schnurr I (ed) (1980) *Idiotypes, what they said at the time. A discussion at Les-Baux-de-Provence, 2nd and 3rd April, 1976*. Editiones <Roche>, Basel
7   Koprowski H (1985) Unconventional vaccines: immunization with anti-idiotype antibody against viral diseases. *Cancer Res* 45: 4689–4690

References and further reading, Part III

8 Herlyn D, Ross AH, Iliopoulos D, Koprowski H (1987) Induction of specific immunity to human colon carcinoma by anti-idiotypic antibodies to monoclonal antibody CO17-1A. *Eur J Immunol* 17: 1649–1652

9 Schnurr I (ed) (1982) *Idiotypes, antigens on the inside.* Editiones <Roche>, Basel

10 Coutinho A (1980) The self-nonself discrimination and the nature and acquisition of the antibody repertoire. *Ann Immunol (Paris)* 131D: 235–253

11 Keating P, Cambrosio A (1997) Helpers and suppressors: On fictional characters in immunology. *J Hist Biol* 30: 381

12 Green DR, Webb DR (1993) Saying the 'S' word in public. *Immunol Today* 14: 523

13 Golub ES, Green DR (1991) *Immunology, a sysnthesis.* Sinauer, Sunderland, Mass.

Further reading

Sercarz E, Oki A, Gammon G (1989) Central *versus* peripheral tolerance: Clonal inactivation *versus* suppressor T cells, the second half of the 'thirty years war'. *Immunol Suppl* 2: 9

Chapter 18

References

1 Miller DS (1949) Hume's Deathblow to Deductivism. *The Journal of Philosophy*, Vol. XLVI, No.23
Stove DC, Hume D (1965) Probability, and Induction. *The Philosophical Review* Vol. LXXIV, No.2

2 Schilpp PA (ed) (1974) *The Philosophy of Karl Popper.* LaSalle, Illinois, USA

3 Laudan L (1996) *Beyond Positivism and Relativism. Theory, Method, and Evidence.* Westview Press, Oxford, UK

4 Popper KR (1959) *The Logic of Scientific Discovery.* Hutchinson, London

5 Hume D (1748) *An Enquiry Concerning Human Understanding.* Ed. Selby-Bigge, Oxford University Press, 1893

6 Kant I (1781) *Die Kritik der reinen Vernunft. Prolegomena. Drittes Hauptstück. Metaphysische Anfangsgründe der Mechanik.* Avalable at: www.ikp.uni-bonn.de/kant/aa04/

7 Guyer P (1998, 2004) Kant, Immanuel. In: Craig E (ed) *Routledge Encyclopedia of Philosophy.* Routledge, London. Available at: http://www.rep.routledge.com/article/DB047

8 Fujimura J (1998) Authorizing Knowledge in Science & Anthropology: Comparison with 19th Century Debate on Euclid. *American Anthropologist* Vol. 100, No. 2

9 Pancer Z, Saha NR, Kasumatsu J, Suzuki T, Amemiya CT, Kasahara M, Cooper M (2005) Variable lymphocyte receptors in hagfish. *Proc Nat Acad Sci USA* 102: 9224

10 Crick FH, Barnett L, Brenner S, Watts-Tobin RJ (1961) General nature of the genetic code for proteins. *Nature* 192: 1227–1232

11 Alberts B, Johnson A, Lewis J, Raff M, Roberts K, Walter P (2002) *Molecular Biology of the Cell* (4th ed). Garland Publishing, New York
12 Codon Usage Database. Available at: www.kazusa.or.jp/codon/
13 Dobzhansky T (1973) Nothing in Biology Makes Sense Except in the Light of Evolution. *The American Biology Teacher* 35: 125–129
14 Dreyer WJ, Bennett JC (1965) The molecular basis of antibody formation: a paradox. *Proc Natl Acad Sci USA* 54: 864–869
15 Tonegawa S, Maxam AM, Tizard R, Bernard O, Gilbert W (1978) Sequence of a mouse germ-line gene for a variable region of an immunoglobulin light chain. *Proc Natl Acad Sci USA* 75: 1485–1489
16 Heisenberg W (1927) Über den anschaulichen Inhalt der quantentheoretischen Kinematik und Mechanik. *Zeitschr f Physik* 43: 197
17 Weizsäcker CF (1958) *Zum Weltbild der Physik*. Hirzel, Stuttgart
18 Rescher N (1979) *Leibniz. An Introduction to his Philosophy*. Basil Blackwell, Oxford
19 Eigen M (1993) The origin of genetic information: Viruses as models. *Gene* 135: 37–47
20 Eigen M (1993) Virus strains as models of molecular evolution. (The fifth Paul Ehrlich Lecture.) *Med Res Rev* 13: 385–398
21 Callaghan A (2006) Emergent Properties of the Human Immune Response to HIV Infection: Results from Multi-Agent Computer Simulations. Available at: http://www.ercim.org/publication/Ercim_News/enw64/callaghan.html
22 Lazar C, Kluczyk A, Kiyota T, Konishi Y (2004) Drug Evolution Concept in Drug Design: 1. Hybridization Method. *J Med Chem* 47: 6973–6982
23 McCord, Adams M (1987) *William Ockham*. 2 vols., Notre Dame, Ind.: University of Notre Dame Press, 2nd rev. ed., 1989

Further reading

Popper KR (1963) *Conjectures and Refutations*. Routledge and Kegan Paul, London
Dawkins R (1976) *The Selfish Gene*. Oxford University Press
Darwin C (1859) *On the origin of species by means of natural selection or the preservation of favoured races in the struggle for life*. Available at: http://en.wikisource.org/wiki/The_Origin_of_Species
Dobzhansky Th (1937, 1941, 1951) *Genetics and the Origin of Species*. Columbia University Press, New York

Appendix

Appendix

Repeatedly mentioned or quoted individuals

Avery Oswald T., 1877–1955. Born in Canada. MD Columbia University, New York. Research position Rockefeller University, New York. Important studies on the immunochemistry of the pneumococcus, the causative agent of pneumococcal pneumonia. 1944, at age 67, discovery of DNA as the material that causes pneumococcal transformation, the first indication that DNA is the substance which genes consist of (see chapter 2, 16). About Avery: McCarty M (1985) *The transforming principle – discovery that genes are made of DNA*. W.W. Norton Comp

Baltimore David. Born in New York, USA. PhD Rockefeller University, New York, postdoctoral training Massachusetts Institute of Technology, Albert Einstein College. Positions at Rockefeller University New York, California Institute of Technology, Pasadena. Groundbreaking studies in genetics and tumor virology. Discovered reverse transcriptase, simultaneously with but independently of Howard M. Temin (see chapter 2). Shared Nobel Prize in Physiology or Medicine 1975, with Renato Dulbecco and Howard Temin. Involved in the Imanishi/Baltimore affair concerning alleged scientific fraud in idiotypic network research.

Behring Emil von, 1854–1917. Doctoral degree Berlin, assistantship Robert Koch Institute Berlin, then professorships in Halle and Marburg. Founded Behringwerke in Marburg for commercial production of antitoxins. Discovered diphtheria and tetanus antitoxins, together with S. Kitasato, in 1890 (see chapters 6, 16). Biography: Zeiss H, Biehling R (1941) *Behring, Gestalt und Werk*. Bruno Schultz Verlag, Berlin-Grunewald.

Benacerraf Baruj. Born in Caracas. Medical studies in New York, Virginia, and Paris. Positions at New York University, NIH, Harvard University.

Shared 1980 Nobel Prize in Physiology or Medicine with Jean Dausset and George Snell, for "discoveries concerning genetically determined structures on the cell surface that regulate immunological reactions". Numerous contributions including studies on MHC, Ir genes, suppressor T cells, and antigen-specific T cell factors (see chapters 7, 10, 11).

Bloor David. Trained in philosophy, mathematics and psychology, Cambridge and Edinburgh. PhD in psychology. Published on the Kuhn/Popper debate, the cognitive functions of metaphor, and on the sociology of scientific knowledge (see chapter 2). Author of multiple books, including: Bloor D (1976, 2nd edition 1991) *Knowledge and Social Imagery*. Routledge, Chicago University Press; Bloor D (1996) *Scientific Knowledge: A Sociological Analysis*. Athlone and Chicago University Press.

Bordet Jules J.B.V., 1870–1961. Born in Belgium. Medical Doctor University of Brussels. Assistant with Ilya I. Metchnikov at Pasteur Institute, Paris. Professorships in Paris and Brussels. Discovered hemolysis and complement fixation (with Gengou), quarrels with Paul Ehrlich thereabout (see chapter 16). Nobel Prize in Physiology or Medicine 1919, for "his discoveries relating to immunity". Biography: de Kruif P (1932) *Men against death*. Harcourt, Brace, New York.

Bona Constantin A. Born in Romania. Positions held at the Pasteur Institute Paris, NIH, Bethesda, Mount Sinai Hospital, New York. Important contributions to idiotypic network research by working on regulation of idiotypes associated with antibodies to levan, showed that injection of anti-idiotypic antibody into newborn mice can stimulate production of idiotype (see chapter 14, interview).

Boyse Edward A., 1923–2006. Born in London, Medical Doctor University of London. Position at Sloan-Kettering Institute for Cancer Research, New York. Pioneering work in mouse immunogenetics and hematopoietic development. Important contributions to immunology, together with Harvey Cantor, in the use of antibodies to cell surface markers for lymphocyte subsetting and development (see chapter 11).

Burnet F. Macfarlane, 1899–1985. Born in Australia. Worked at Walter and Eliza Hall Institute. Developed clonal selection theory of antibody formation (see chapters 6, 7, 16). Important contributions to virology. Shared Nobel Prize in Physiology or Medicine with Peter Medawar in 1960, for the "discovery of acquired immunological tolerance". Autobiography: Burnet FM (1968) *Changing Patterns*. Heinemann, Melbourne. Multiple books, including: Burnet FM (1968) *Biology and the Appreciation of Life*. Sun Books, Melbourne; Burnet FM (1974) *Endurance of Life: The Implications of Genetics for Human Life*. Melbourne University Press, Melbourne.

Repeatedly mentioned or quoted individuals

Cantor Harvey. MD New York University. Positions held at National Institutes of Health, National Institute for Medical Research in London, Stanford University, Harvard University. Important contributions to lymphocyte subsetting together with Edward Boyse. Contributed to suppressor T cell circuitry together with Richard Gershon (see chapter 11).

Cazenave Pierre A. French immunologist. Position at Pasteur Institute, Paris. Reported, with J. Oudin, on observation that idiotypes may be shared between antibodies of different specificities, and studies on internal image with N. Jerne (see chapter 9, and 14, interview). Contributions to evolution of light chain genes, studies on malaria. Book: Kohler H, Urbain J, Cazenave PA (eds) (1984) *Idiotypy in Biology and Medicine*. Academic Press, New York.

Claman Henry N. MD New York University. Positions at Barnes Hospital, St. Louis, Massachusetts General Hospital, Boston, University of Colorado, Denver. His early observation on bone marrow-thymus mixtures in adoptive transfer experiments were the first indications of T cell-B cell cooperation in the immune response (see chapter 7). About Claman: Dreskin S (2006) Henry Claman in profile: Identification of cellular cooperation in antibody production. *Journal of Allergy and Clinical Immunology* 117(4): 959–960.

Cohn Melvin. Studied biochemistry, PhD New York University. Postdoctoral fellow, Pasteur Institute, Paris, France. Since 1960 at Salk Institute, San Diego. Numerous contributions to problems of immune recognition, immune response, and antibody diversity. Together with Peter Bretscher author of the associative recognition theory of acquired immunity, and most prominent critic of the network theory (see chapter 7, 9).

Coutinho Antonio. Born in Portugal. MD from Lisbon Medical School. PhD Karolinska Institute in Stockholm. Positions held at Umeå University Medical School, Basel Institute for Immunology, Pasteur Institute in Paris, Gulbenkian Science Institute, Portugal. Important contributions to the question of tolerance and self-nonself discrimination in immune networks. Theoretical and conceptual contributions to network theory, in part together with Francesco Varela (see chapters 12 and 14, interview).

Crick Francis H., 1916–2004. Born in Britain. Trained in physics, biochemistry, and biology. Developed, together with James Watson, the double helix model of DNA. Shared the Nobel Prize in Physiology or Medicine with Watson and M.H.F. Wilkins in 1962, for "discoveries concerning the molecular structure of nucleic acids and its significance for information transfer in living material". Important contributions to protein synthesis and genetic code, put forward the central dogma of biology (see chapter 2, 18). Autobiography: Crick

FH (1988) *What Mad Pursuit: A Personal View of Scientific Discovery*. Basic Books, New York.

Davis Mark M. PhD in molecular biology, California Institute of Technology. Positions held at NIH and Stanford University. First to identify a T cell receptor gene, simultaneously with T. Mak (see chapter 10). Multiple contributions to the molecular characterization of T cell antigen recognition.

Dobzhansky Theodosius G., 1900–1975. Born in Ukraine. Trained in genetics in Kiew and St. Petersburg. Emigrated to USA, positions at Columbia University, New York, California Institute of Technology, Pasadena, and Rockefeller University, New York. Important contributions to evolution (see chapter 18). Multiple books, including: Dobzhansky TG (1937, 2nd ed. 1941; 3rd ed. 1951) *Genetics and the Origin of Species*. Columbia University Press, New York; Dobzhansky TG (1955) *Evolution, Genetics, & Man*. Wiley & Sons, New York; Dobzhansky TG (1962) *Mankind Evolving*. Yale University Press, New Haven, Connecticut.

Doherty Peter C. Born in Australia. Studied veterinary medicine, PhD University of Edinburgh. Positions held at Australian National University in Canberra, University of Philadelphia, USA, and University of Memphis, USA. Discovered, together with R. Zinkernagel, MHC-restricted recognition of antigens by T cells (see chapters 7, 10). Shared with Zinkernagel the 1996 Nobel Prize in Physiology or Medicine for "their discoveries concerning the specificity of cell-mediated defense". Multiple contributions to antiviral immunity.

Eco Umberto. Born in Italy. Studied philosophy and history of literature, doctoral degree University of Turin. Worked as culture critic for print media and television. Professorial position at University of Bologna in linguistics and semiotics. Multiple guest professorships and honorary degrees at universities worldwide. Co-organized a conference on immunology and semiotics in 1984 (see chapter 12). The British magazine *Prospect* elected him 2005 as second most important intellectual of the world, after Noam Chomsky and before Richard Dawkins. Published multiple scientific books, including: Eco U (1976) *A Theory of Semiotics*. Indiana University Press, Bloomington; Eco U (1984) *Semiotics and the Philosophy of Language*. McMillan Press, Houndsmill; and fiction, including Eco U (1983) *The Name of the Rose*. Harcourt, New York; Eco U (1989) *Foucault's Pendulum*. Secker & Warburg, London.

Edelman Gerald M. Studied medicine. MD University of Pennsylvania. Medical positions Massachusetts General Hospital, Boston, and American Hospital, Paris. PhD Rockefeller University, New York. Positions held at

Rockefeller University and Scripps Clinic and Research Institute, La Jolla. Discovered that antibodies consist of heavy and light chains, first amino acid sequence of a complete IgG molecule (see chapters 6, 16). Important contributions to neuroscience. Shared with R.R. Porter the 1972 Nobel Prize in Physiology or Medicine for "their discoveries concerning the chemical structure of antibodies". Books include: Edelman GM (1987) *Neural Darwinism: The Theory of Neuronal Group Selection*. Basic Books, New York; Edelman GM (1993) *Topobiology: An Introduction to Molecular Embryology*. Basic Books, New York.

Ehrlich Paul, 1854–1915. Born in Silesia. Studied Medicine. Doctoral degree University of Leipzig. Positions held in Breslau, Berlin, Frankfurt. Important contributions to the properties of tissue dyes, discovered mast cells, multiple groundbreaking studies on antibodies, developed the first antibiotic drug for syphilis, salvarsan. Put forward the side chain theory of antibody formation (see chapters 6, 16). Shared with Ilya I. Metchnikov the 1908 Nobel Prize in Physiology or Medicine "in recognition of their work on immunity". Biographies: Bäumler E (1984) *Paul Ehrlich, Scientist for Life*. Holmes and Meier, New York and London; Marquardt M (1951) *Paul Ehrlich*. Springer Verlag, Berlin.

Feyerabend Paul, 1924–1994. Born in Austria, studied history, sociology, and philosophy. Joined Karl Popper at the London School of Economics, by whom he was first influenced but whose philosophy he later rejected, developing his own "scientific anarchism". Positions held in Bristol, GB, Minnesota and Berkeley, USA, Zurich, Switzerland. Published multiple books with severe attacks against scientific methodology (see chapter 2), including: Feyerabend P (1975) *Against Method*. Verso, London; Feyerabend P (1978) *Science in a Free Society*. New Left Books, London; Feyerabend P (1987) *Farewell to Reason*. Verso/New Left Books, London.

Fleck Ludwik, 1896–1961. Born in Poland. Studied medicine, doctoral degree University of Lvov. Positions as medical microbiologist held in Lvov University and General Hospital. Developed typhoid vaccine in Warsaw's Jewish ghetto, was forced to oversee typhoid vaccine production in Auschwitz and Buchenwald concentration camps. Professorial positions in Lublin and Warsaw. Developed doctrine on thought style and thought collective in the generation of scientific facts (see chapter 4). Major book: Trenn TJ, Merton RK (eds; foreword by Thomas Kuhn) (1979) *The Genesis and Development of a Scientific Fact*. University of Chicago Press, Chicago. About Fleck: Cohen RS, Schnelle T (1986) *Cognition and Fact – Materials on Ludwik Fleck*. Reidel, Dordrecht.

Fujimura Joan. PhD in sociology, University of Berkeley, California. Positions held at Harvard University, Stanford University, University of

Wisconsin. Research subjects: Anthropology, sociology, cultural studies, feminist studies, and history of science, medicine, and technology (see chapter 5, 17). Book: Fujimura J (1999) *Crafting Science: A Socio-History of the Quest for the Genetics of Cancer.* Harvard University Press, Boston.

Gershon Richard K., 1932–1983. Studied medicine, MD Yale University. Trained in pathology, Yale University. Pre- and postdoctoral scientific experiences in France and Japan. Position at Yale University. Research on viral hepatitis. First to describe suppressor T cells, multiple contributions on suppressor T cell circuits, contrasuppressor T cells, suppressor T cells factors, and VH- and I-J-restricted interactions of suppressor T cells (see chapters 7, 11).

Germain, Ronald N. American immunologist, postdoctoral training with B. Benacerraf, Harvard University, Boston. Multiple contributions to suppressor T cell circuits and factors (see chapter 11, and 14, interview). Working at NIH, Bethesda, important contributions on T cell antigen recognition and development.

Gowans James L. Born in England. PhD Oxford University. Position held at Medical Research Council, Oxford. Major contribution to the identification of lymphocytes as immunological effector cells and their recirculation (see chapter 6). Major papers: Gowans JL (1962) The fate of parental strain small lymphocytes in F 1 hybrid rats. *Ann NYAcad Sci* 99: 432–455; Gowans JL, McGregor DD, Cowen DM (1962) Initiation of Immune Responses by Small Lymphocytes. *Nature* 196: 651.

Greene Mark I. Born in Canada. MD and PhD University of Manitoba. Positions held at Harvard University, Boston, and Penn State University, Pennsylvania, and Oxford University. Numerous contributions to idiotypic research, suppressor T cell factors and circuits (see chapter 10), more recently to oncogene research. Book edited: Greene MI, Nisonoff A (1984) *The Biology of Idiotypes.* Plenum Press, New York, London.

Hacking Ian. Born in Canada. Studied philosophy at University of British Columbia and University of Cambridge, GB. PhD University of Cambridge. Positions held at University of Makarere, Uganda, University of Cambridge, Stanford University, USA, and University of California, Santa Cruz. Hacking combines "entity realism", i.e. a scientific entity such as "the electron is real because human beings use it for manipulations", with a profound scepticism towards scientific laws (see chapter 4). Multiple books, including: Hacking I (1999) *The Social Construction of What?* Harvard University Press, Cambridge; Hacking I (1995) *Rewriting the Soul: Multiple Personality and the Sciences of Memory.* Princeton University Press; Hacking I (1990) *The Taming of Chance.* Cambridge University Press.

Repeatedly mentioned or quoted individuals 255

Haurowitz Felix, 1896–1987. Born in Czechoslovakia. Studied Biochemistry in Prague and Munich. Doctoral degree University of Prague. Positions held at Universities of Prague and Heidelberg, University of Istanbul, Turkey, and Indiana University, Bloomingdale, USA. Important contributions to structural analyses of hemoglobin in cooperation with his cousin-in-law Max Perutz. Proposed, together with virologist Friedrich Breinl, the earliest version of an instructional theory of antibody specificity (see chapters 6, 16).

Häyry Pekka. Finnish immunologist, works at University of Helsinki. Multiple contributions to transplantation immunology and medicine. Discovered cytotoxic T cells (see chapter 7).

Heidelberger Michael, 1888–1991. Studied chemistry and obtained PhD at Columbia University, New York. Postdoctoral education in Zurich, Switzerland. Positions at Rockefeller University, New York, Columbia University, New York, Rutgers University, and New York University. Pioneering work, together with O.T. Avery, on the immunochemistry of pneumococcus, contributed quantitative procedures for determination of antibodies, purification of antibodies (see chapter 16). Autobiographical articles: Heidelberger M (1977) A "Pure" Organic Chemist's Downward Path. *Ann Rev Microbiol* 31: 1–12; Heidelberger M (1984) Reminiscences. *Immunological Reviews* 82: 7–27. About Heidelberger: Cruse JM (1988) A Centenary Tribute: Michael Heidelberger and the Metamorphosis of Immunologic Science. *Journal of Immunology* 140: 2861–2863.

Herzenberg Leonard N. PhD in Biochemistry and Immunology, California Institute of Technology, Pasadena. Postdoctoral fellow Pasteur Institute, Paris. Professorial position at Stanford University. Important contributions to mouse allotypy, described allotype-specific suppressor T cells (see chapter 9, 11). Pioneered the development of the fluorescence-activated cell sorter (FACS) and among the first to produce monoclonal antibodies to lymphocyte surface antigens.

Hilschmann Norbert. Born in Germany. Studied Biochemistry, PhD University of Munich. Postdoctoral training Rockefeller University, New York. Position at Max-Planck-Institute of Experimental Medicine in Göttingen. First to determine complete amino acid sequences of two Bence Jones proteins, discovered variable and constant regions of immunoglobulin light chains, with L. Craig (see chapters 6, 16). Important contributions to structure of MHC antigens.

Hoffmann Geoffrey W. Born in Australia. Studied mathematics. Postdoctoral training at Max Planck Institut für Biophysikalische Chemie, Göttingen, Basel Institute for Immunology. Positions held at University of

British Columbia, Los Alamos National Laboratory. Developed a theory of the regulation of the immune system called the symmetrical network theory (see chapter 9). Continues to study immune networks in relation to HIV and vaccination. Paper: Hoffmann GW (2004) Proteomic analyser with applications to diagnostics and vaccines. *Journal of Theoretical Biology* 228: 459–465.

Hood Leroy. MD John Hopkins University, Baltimore, PhD California Institute of Technology, Pasadena. Positions held at California Institute of Technology, University of Washington and Institute of System Biology, Seattle, Washington. Pioneered development of apparatus for rapid automated protein and nucleic acid synthesis and sequence analysis. Multiple contributions to molecular immunology and genetics. His group demonstrated absence of a gene for I-J, and lack of expression of immunoglobulin genes in T cells (see chapters 10, 11).

Hübner Kurt. German philosopher. Professorship University of Kiel. Proposes that science is as conditional as any other ontology, including mythology and religion (see chapter 2). Multiple books, including: Hübner K (2003) *Das Christentum im Wettstreit der Weltreligionen. Zur Frage der Toleranz.* Mohr Siebeck Verlag, Tübingen; Hübner K (2001) *Glaube und Denken. Dimensionen der Wirklichkeit.* Mohr Siebeck Verlag, Tübingen.

Hume David, 1711–1776. Born in Scotland. Studied jurisprudence. Worked in Edinburgh, Paris, Vienna, as librarian, secretary, and writer. Reserved axiomatic cognition to mathematics and geometry, any other knowledge is acquired by committing frequently encountered connections to memory, mostly without understanding the reasons behind (see chapter 2, 18). Writings include: Hume D (1739–1740) *Treatise on human nature*; Hume D (1748) *Enquiry concerning human understanding*; Hume D (1751) *Enquiry concerning the principles of morals*; Hume D (1755) *The natural history of religion*.

Janeway Charles A. Jr., 1943–2003. Studied medicine, MD Harvard University, Boston. Postdoctoral work at Harvard University, National Institute for Medical Research, London, University of Cambridge, England, University of Uppsala, Sweden. Positions held at NIH, Bethesda, Yale University, New Haven. Multiple contributions to basic immunology. Debate with P. Matzinger on danger signals and innate immunity (see chapter 13). Coautor of the widely used textbook: Janeway CA Jr (6th ed. 2006) *Immunobiology: The Immune System in Health and Disease.* Autobiographical article: Janeway CA Jr (2002) A trip through my life with an immunological theme. *Annual Review of Immunology* 20: 1–28.

Joao Cristina. Portugese immunologist, works at Gulbenkian Institute, Oeiras, Portugal. Recently published papers, together with M. Cascalho, describing a strongly reduced T cell repertoire diversity in mice that do not express diverse immunoglobulins (see chapters 13, 14, interviews Urbain, Rajewsky).

Jerne Niels, 1911–1994. Born in England to Danish parents. Studied medicine, MD University of Copenhagen. Positions at Danish National Serum Institute, University of Pittsburg, USA, World Health Organization, Geneva, Paul Ehrlich Institute, Frankfurt, Basel Institute of Immunology, Basel. Proposed the natural selection theory of antibody formation and the idiotypic network theory (see chapters 6, 9, 17). Shared the 1984 Nobel Prize in Physiology or Medicine with C. Milstein and G.J.F. Köhler for "theories concerning the specificity in development and control of the immune system...". Biography: Söderquist T (2003) *Science as autobiography, the troubled life of Niels Jerne*. Yale University Press, New Haven, London.

Kabat Elvin A., 1914–2000. PhD Columbia University, New York. Postdoctoral training Svedberg Laboratories, Uppsala, Sweden. Discovered distinction of 19S and 7S antibodies, and association of antibodies with γ-globulin serum fraction, with K. Pedersen and A. Tiselius (see chapter 16). Positions at New York University, NIH, Bethesda. Numerous contributions to immunochemistry, including chemical nature of blood group antigens. Several widely used textbooks, including: Kabat EA (1961 2nd ed.) *Kabat and Mayer's Experimental Immunochemistry*. Illustrated by Chas CT, Springfield; Kabat EA (1976) *Structural Concepts in Immunology and Immunochemistry*. Holt, Rinehart, and Winston, New York. Autobiographical article: Kabat EA (1988) Before and after. *Annu Rev Immunol* 6: 1–24.

Kazatchkine Michel D. Born in France. Studied medicine in Paris, trained in immunology at Pasteur Institute, St. Mary's Hospital, London, Harvard Medical School, Boston. Important contributions to intravenous immunoglobulin therapy of autoimmune diseases, proposed that the beneficial effects are due to idiotypic network reconstitution (see chapter 13).

Kitasato Shibasaburo, 1853–1931. Japanese physician and bacteriologist, trained with R. Koch and A.v. Wassermann in Berlin, worked with E. v. Behring on diphtheria and tetanus antitoxin therapy (see chapters 6, 16). Founder of the Kitasato Institute in Japan. Important contributions to pathogenesis of tuberculosis and bubonic plaque. Biography: Kyle RA (1999) Shibasaburo Kitasato-Japanese bacteriologist. *Mayo Clinic Proceedings* 74: 146

258   Appendix

Knorr-Cetina Karin. Sociologist. Professorial positions at University of Konstanz and Institute for World-Society Studies, University of Bielefeld. Multiple contributions to the sociology of science, including: Knorr Cetina KD (1998) Constructivism. In: Davis JB, Hands DW, Mäki U (eds) *Handbook of Economic Methodology*. E. Elgar Publ., Cheltenham; Knorr Cetina KD (1997) What Scientists Do. In: Ibanez T, Iniquez L (eds) *Critical Social Psychology*. Sage, London; Knorr Cetina KD (1995) *The Care of the Self and Blind Variation: An Ethnography of the Empirical in Two Sciences*. In: Galison P, Stump D (eds) *The Disunity of Science. Boundaries, Contexts, and Power*. Stanford University Press, Stanford (see chapters 4, 15).

Köhler, Georges J.F., 1946–1995. German immunologist. Studied biology. PhD Basel Institute of Immunology and University of Freiburg. Postdoctoral training MRC Cambridge, BG, and Basel Institute of Immunology. Research position at Max-Planck-Institute of Immunobiology, Freiburg. Invented hybridoma technique, together with C. Milstein (see chapter 16). Shared 1984 Nobel Prize in Physiology or Medicine with C. Milstein and N. Jerne, for "...the discovery of the principle for producing monoclonal antibodies". Biography: Eichmann K (2005) *Köhler's Invention*. Birkhäuser Verlag, Basel.

Köhler Heinz. German immunologist. PhD University of Munich, postdoctoral training Max-Planck-Institute of Biochemistry, Munich. Research positions at University of Kentucky, Lexington, Kentucky, IDEC Pharmaceuticals Corp. La Jolla, California, San Diego Regional Cancer Centre in California. Early contributions to idiotypic regulation, first to report detection of spontaneous auto-antiidiotypic antibodies (see chapter 9, and 14, interview). Research towards antibody-based therapies of cancer and AIDS.

Koprowski Hilary. Polish virologist and immunologist. MD University of Warsaw. Position at the Wistar Institute of Science, Philadelphia, USA. Developed vaccines for polio and rabies. Early production of monoclonal antibodies for virus typing and to cancer antigens. Widely publicized attempts towards idiotypic vaccination against colon cancer (see chapter 17).

Kuhn Thomas S., 1922–1996. Studied physics, PhD Harvard University, Boston. Professorial positions in history of physics, Harvard University, in philosophy of science, University of Berkeley, California, Princeton University, and Massachusetts Institute of Technology. Proposed discontinuous development of science as a repeating sequence of normal science, crisis, revolution, and paradigm shift. Most important book: Kuhn TS (1962) *The Structure of Scientific Revolutions*. University of Chicago Press. About Kuhn: Bird A (2000) *Thomas Kuhn*. Princeton University Press and

Acumen Press; Fuller S (2000) *Thomas Kuhn: A Philosophical History for Our Times*. University of Chicago Press.

Kunkel Henry G., 1916–1983. MD Johns Hopkins University, Baltimore. Medical position at Bellevue Hospital, New York. Position at Rockefeller University and Hospital, New York. Demonstrated that myeloma proteins are immunoglobulins. Multiple contributions to immunochemistry of immunoglobulins, and immunological disorders. Discovered individual antigenic specificity of myeloma proteins, equivalent to idiotypy of antibodies (see chapters 6, 8).

Lakatos Imre, 1922–1974. Born in Hungary. Studied physics and philosophy, doctoral degree University of Debrecen, Hungary. Continued studies at Moscow University, then worked as state officer in the ministry of education in Budapest. Doctoral degree in philosophy University Cambridge, GB. Research position at London School of Economics, dispute and disagreements with Karl Popper (see chapter 3). His major book: Lakatos I (1976) *Proofs and Refutations*. Cambridge University Press, was published after his death.

Landsteiner Karl, 1868–1943. Born in Austria. Studied medicine, doctoral degree from University of Vienna. Postdoctoral work in Zürich, Würzburg, and Munich. Positions in Vienna, den Haag, the Netherlands, Rockefeller University, New York. Multiple discoveries, including human blood groups and rhesus factor, infectious origin of poliomyelitis, first auto-antibody in paroxysmal hemoglobinuria, anti-hapten antibodies (see chapters 6, 16). Nobel Prize in Physiology or Medicine 1930, for "his discovery of human blood groups".

Latour Bruno. French philosopher and anthropologist. Professorial position at the Ecole Nationale Supérieure des Mines in Paris and, for various periods, visiting professor at University of California San Diego, at the London School of Economics, and Harvard University. Social constructionist approach to science. Undertook an ethnographic study of a neuroendocrinology research laboratory at the Salk Institute (see chapter 4), published with co-author Steve Woolgar under the title: Latour B, Woolgar S (1979) *Laboratory Life: the Social Construction of Scientific Facts*. Sage, Los Angeles, USA. More recently involved in actor-network theory. Further books include: Latour B (1993) *We have never been modern*. Harvard University Press, Cambridge Mass., USA; Latour B (1996) *Aramis, or the love of technology*. Harvard University Press, Cambridge Mass., USA; Latour B (1999) *Pandora's hope: essays on the reality of science studies*. Harvard University Press, Cambridge Mass., USA.

Laudan Larry. American science philosopher. PhD Princeton University. Professorial positions at University College, University of London, Cornell University, New York, University of Pittsburgh, multiple visiting professorships including London School of Economics, University of Illinois-Chicago. Maintains that all constructionist science theories, particularly underdeterminism and relativism, have been falsified (see chapter 3). Publications include: Laudan L (1996) *Beyond Positivism and Relativism.* Westview Press, Boulder; Laudan L (1990) *Science and Relativism: Dialogues on the Philosophy of Science.* University of Chicago Press, Chicago; Laudan L (1990) De-Mystifying Underdetermination. In: Savage W (ed) (1990) *Scientific Theories.* University of Minnesota Press, Minneapolis.

Lederberg Joshua, 1925–2008. Studied medicine and biology, PhD Yale University, New Haven. Positions at University of Madison, Wisconsin, Stanford University, Rockefeller University. Multiple contributions to molecular biology, discovered sexual reproduction in bacteria, transfer of genetic information by bacteriophage. Shared 1958 Nobel Prize in Physiology or Medicine with G.W. Beadle and E.L. Tatum for "discoveries concerning genetic recombination and the organization of the genetic material of bacteria". One of the first to acknowledge clonal selection theory of antibody formation, contributed theory on self-tolerance (see chapter 6).

Levy Ronald. Studied medicine, MD Stanford University. Postdoctoral training NIH, Bethesda, Weizmann Institute of Science, Israel. Position at Stanford University. First to implement procedures to treat human B cell lymphoma with anti-idiotypic antibodies (see chapter 13).

Mak Tak W. Born in China. PhD University of Alberta, Canada. Positions at Ontario Cancer Institute, Toronto, University of Toronto, Amgen Research Institute, Toronto. First to discover a T cell receptor gene, simultaneous with M.M. Davis (see chapter 10). Multiple contributions to molecular immunology, in part by production of knockout mice.

Marrack Philippa C. Born in Britain. Studied biochemistry at Cambridge University, GB. Postdoctoral training at MRC, Cambridge, University of California San Diego. Positions at University of Rochester, and National Jewish Medical and Research Center, Denver, Colorado. Multiple contributions to T cell antigen recognition and development, superantigens. One of the first to produce monoclonal antibodies to T cell receptor (see chapter 10).

Marchalonis John J., 1941–2007. Studied biology, PhD Rockefeller University, New York. Postdoctoral training at Walter and Eliza Hall

Institute, Melbourne. Research positions at Brown University, Rhode Island, Walter and Eliza Hall Institute, NCI, Frederick, MD, Medical University of South Carolina, University of Arizona. Multiple contributions on phylogeny of immunoglobulin structure, controversial studies on the immunoglobulin nature of the T cell receptor (see chapter 10).

Matzinger Polly C. E. PhD University of California, San Diego, postdoctoral training University of Cambridge and Basel Institute of Immunology. Position at NIH, Bethesda. Multiple contributions, including influential theoretical papers, to T cell responsiveness and tolerance. Proposed the danger model of the immune system (see chapter 13).

McDevitt Hugh O. Studied medicine, MD Harvard University, medical training Boston, New York, Japan. Postdoctoral training Harvard Medical School, National Institute of Medical Research, Mill Hill, GB. Position at Stanford University. Multiple contributions to role of MHC in the immune system, first to describe the I region of H-2 complex. Discovered, together with Michael Sela, the genetic control of immune response by Ir genes, proposed together with B. Benacerraf that T cell receptor is encoded by MHC-linked genes (see chapter 7, 10).

Medawar Peter B., 1915–1987. Born in Brazil. Studied zoology in Oxford. Positions at University of Birmingham, University College, London, National Institute for Medical Research, London, Royal Postgraduate Medical School. Demonstrated induced tolerance to tissue grafts in mice pretreated as newborns with donor cells (see chapters 6). Shared with F. M. Burnet the 1960 Nobel Prize in Physiology or Medicine for "the discovery of acquired immunological tolerance." Multiple philosophical books, including: Medawar PB (1959) *The Future of Man*. Methuen, London; Medawar PB (1967) *The Art of the Soluble*. Methuen, London; Medawar PB (1972) *The Hope of Progress*. Methuen, London; Autobiography: Medawar PB (1986) *Memoir of a Thinking Radish*. Oxford University Press, Oxford.

Miller Jaques F.A. Born in France. Studied medicine at University of Sydney, Australia. Postdoctoral training at Royal Prince Alfred Hospital, Sydney, University College, London, GB. Research positions at NIH, Bethesda, and Walter and Eliza Hall Institute, Melbourne. Discovered, together with G. Mitchel, role of the thymus in T cell development, and T cell-B cell cooperation. Multiple contributions to cell-mediated immunity and self-tolerance (see chapter 7).

Milstein Cesar, 1927–2002. Born in Argentina. Studied biochemistry, doctoral degree University of Buenos Aires. Research positions at University of Buenos Aires, University of Cambridge, GB, and MRC, Cambridge.

Multiple contributions to antibody structure and genetics, somatic mutation. Invented, together with G.J.F. Köhler, the technique for producing monoclonal antibodies (see chapter 16). Shared 1984 Nobel Prize in Physiology or Medicine with G.J.F. Köhler and N.K. Jerne for "... the discovery of the principle for producing monoclonal antibodies".

Mitchison N. Avrion. British immunologist. Positions at University College, London, Deutsches Rheumaforschungszentrum Berlin. First to describe that immunity to transplants is transferred more easily by cells than by serum. Important contributions to T cell-B cell cooperation by adoptive transfer experiments (see chapter 7).

Möller Göran. Swedish immunologist. Research position at University of Stockholm. Important contributions to use of fluorescent antibodies to cell surface antigens in microscopy, B cell activation. Initiated process of discrimination of suppressor T cells by an article, entitled Möller G (1988) Do Suppressor T cells exist? *Scandinavian Journal of Immunology* 27: 247 (see chapters 7, 11).

Nisonoff Alfred, 1923–2001. Studied biochemistry, PhD Johns Hopkins University, Baltimore. Postdoctoral training at Roswell Park Memorial Institute, Buffalo. Research positions at University of Illinois, Chicago, and Brandeis University, Massachusetts. Important contributions to anti-hapten antibodies, generated the anti-arsonate crossreactive idiotypic system. First demonstration of idiotype suppression by injection of anti-idiotypic antibodies, suggested use of internal image anti-idiotypes for vaccination, cooperative work on idiotype-specific suppressor T cells (see chapters 8, 9,11).

Nossal Gustav J.V. Born in Austria. Studied medicine at Sydney University, Australia. Postdoctoral training and research position Walter and Eliza Hall Institute, Melbourne. First to corroborate clonal selection by single-cell assays (see chapter 6). Multiple contributions to immunological specificity, self-nonself discrimination, tolerance, vaccination. Important activities in global health strategies, science and society. Books include: Nossal GJV, Coppel RL (2002) *Reshaping Life: Key Issues in Genetic Engineering.* Cambridge University Press

Ohno Susumu, 1828–2000. Born in Korea of Japanese parents. Studied veterinary science, PhD Tokio and Hokkaido Universities. Research positions at University of California, Los Angeles, and City of Hope Medical Center, Duarte, California. Important contributions to evolution biology by studies on gene duplication. Proposed an essential role of short sequence repeats in evolution, as well as in language and music. Translated DNA sequences into music (see chapters 6, 12). Book: Ohno S (1970) *Evolution by Gene Duplication.* Springer-Verlag, Berlin.

Oudin Jacques, 1908–1985. Studied medicine, MD University of Paris. Position at Institute Pasteur, Paris. Developed important assays of antigen-antibody interaction in agar gels. Discovered immunoglobulin allotypes and idiotypes, the latter providing the most important proto-idea of network theory (see chapter 8).

Paul William E. American immunologist. Studied medicine, positions at New York University, and NIH, Bethesda. Cooperated with C. Bona on idiotypes associated with antibodies to bacterial levan, otherwise distanced relationship to network theory (see chapter 14, interview). Important contributions to research on cytokines including IL-4, Th1/Th2 regulation.

Pauling Linus C., 1901–1994. Studied chemistry, PhD in physical chemistry and mathematical physics, California Institute of Technology. Position at California Institute of Technology. Important contribution to the nature of the chemical bond, electron structure of the atomic nucleus, X-ray crystallography of proteins. Proposed an influential version of instructional theory of antibody specificity, invoking antigen-induced refolding of immunoglobulin chains (see chapter 6, 16). Later engaged in anti-atomic and peace activities. Nobel Prize in Chemistry, 1954; Nobel Peace Prize, 1962.

Popper Karl R., 1902–1994. Born in Vienna. PhD in psychology, teacher at secondary school. Professorial positions at Canterbury University New Zealand, London School of Economics. In his influential book: *The Logic of Scientific Discovery*, (1934 as Logik der Forschung, English translation 1959), he gives an account of scientific method by empirical falsification, thus repudiating the classical justificationist approach to scientific knowledge (see chapter 3). Also commented on political philosophy, liberal democracy, open society. Multiple books, bibliography in: Lube M (2005) *Karl R. Popper. Bibliographie 1925–2004. Wissenschaftstheorie, Sozialphilosophie, Logik, Wahrscheinlichkeitstheorie, Naturwissenschaften.* Peter Lang, Frankfurt/Main etc., 576 pp.

Porter Rodney R., 1917–1985. Born in Britain. PhD in biochemistry, University of Liverpool. Research positions at National Institute of Medical Research, Mill Hill, St. Mary's Hospital Medical School, and University of Oxford. Determined the enzymatic digestion fragments of immunoglobulin, Fab and Fc. Shared the 1972 Nobel Prize in Physiology or Medicine with G.M. Edelman, for "their discoveries concerning the chemical structure of antibodies" (see chapter 6, 16).

Potter Michael. American immunologist. Research position at NIH, Bethesda. Developed the technique for production of plasmacytoma tumors in mice. Generated multiple mouse myelomas, including the one used by C. Milstein and G.J.F. Köhler for production of hybridomas. Many of Potter's myelomas were used as reference idiotypes in crossreactive idiotypic systems (see chapters 6).

Quine Willard v.O., 1908–2000. Studied philosophy and mathematics, PhD Harvard University. Educational visits with Lvov–Warsaw School and Vienna Circle. Professorial position at Harvard University. Formulated "Conformation Holism": All theories are under-determined by empirical data. If a theory fails to fit with the data or is unworkably complex, there are always many equally justifiable alternatives. Agrees with Pierre Duhem that for any collection of empirical evidence there are many theories that are able to account for it (Duhem-Quine Thesis, see chapter 2). About Quine: Gibson RF (1982/1986) *The Philosophy of W.V. Quine: An Expository Essay.* University of South Florida, Tampa.

Raff Martin. Born in Canada. Studied medicine, postdoctoral training at National Institute of Medical Research. Position at University College London. Discovered Thy-I antigen of mouse T cells, first to describe discrimination of T cells and B cell on the basis of surface markers, Thy-I for T cells and immunoglobulin for B cells (see chapters 7, 11). Subsequently engaged in neurobiology.

Rajewsky Klaus. Born in Germany. Studied medicine and chemistry. Trained at University of Cologne, Institute Pasteur, Paris. Positions at University of Cologne and Harvard University, Boston. Among the first to study T cell-B cell cooperation (see chapter 7). Influential experiments on nylon fiber-purified T cell receptors containing VH region, together with K. Eichmann demonstrated induction of T and B cell responses by injecting anti-idiotypic antibodies into mice (see chapters 9, 11, and 14, interview). Multiple contributions to molecular immunology, in part by producing knockout mice.

Reichenbach Hans, 1891–1953. Born in Germany. Studied mathematics, physics, and philosophy. Professorial positions University of Berlin, University of Istanbul, Turkey, University of California Los Angeles. Representative of logical empiricism and positivism, strong ties to Vienna Circle. Tried to develop quantum logic with *uncertain* as a third category in addition to *true* and *false* (see chapters 2, 18). Multiple books, including: Reichenbach (1944) *Philosophic foundations of quantum mechanics.* University of California Press, Berkeley and Los Angeles; Reichenbach H (1951) *The rise of scientific philosophy.* University of California Press, Berkeley.

Rodkey, Scott L. Studied microbiology, PhD University of Kansas. Postdoctoral training in immunology University of Illinois, Chicago. Research positions Kansas State University, California Institute of

Technology, Basel Institute of Immunology, University of Texas, Houston. First to demonstrate the production of auto-anti-idiotypic antibody by reinjection of isolated rabbit antibodies into the same individual (see chapter 9).

Sachs David H. Studied medicine, MD Harvard University. Postdoctoral training Massachusetts General Hospital, and NIH, Bethesda. Positions at NIH, University of Uppsala, Sweden, Massachusetts General Hospital. Multiple contributions to MHC, transplantation immunology, large animal models for transplantation. Developed crossreactive idiotypic system based on mouse antibodies to staphylococcal nuclease. Attempts to counteract transplant rejection by anti-idiotypic antibodies (see chapter 8, and 14, interview).

Sakagushi Shimon. Japanese immunologist, studied medicine, works at University of Kyoto. Important contributions to regulatory T cells, their role in transplantation immunology and inflammation (see chapter 11).

Sela Michael. Born in Poland, position at Weizmann Institute of Science, Israel. Discovered, together with H. McDevitt, genetic control of immune response by Ir genes (see chapter 7). Immunochemical studies on synthetic antigens, random copolymer peptides, including their therapeutic effects in autoimmune disease.

Sercarz Eli E. PhD in immunology Harvard University. Postdoctoral training Harvard University and Massachusetts Institute of Technology. Research Positions at University of California Los Angeles and Torrey Pines Institute for Molecular Studies, San Diego. Important contributions to immunogenicity of proteins, using hen egg lysozyme as model. Multiple contributions to idiotype-specific suppressor T cells, I-J restriction (see chapter 11, and 14, interview). Studies on T cells in autoimmune disease. Coorganizer of Immuno-Semiotics conference with U. Eco (see chapter 13).

Simonsen Morton. Danish immunologist. Developed an assay in which chicken spleen cells are distributed in limiting numbers onto the chorioallantois of allogeneic chicken embryos. Cells that recognize histocompatibility antigens on the chorioallantois proliferate and build colonies which can be enumerated. This assay showed high proportions of alloreactive cells, which were hard to reconcile with clonal selection theory (see chapter 7).

Sokal Alan. American physicist. PhD Princeton University. Positions at University of Nicaragua, New York University, New York, and University College, London. Submitted a purposely non-sensical philosophical paper to *Social Text* which was accepted by the editors and published in an issue that otherwise contained contributions of science critics. When he published, in another journal, that his paper was a hoax, this became known as the Sokal Affair and was the start of Science Wars (see chapter 5).

Tada Tomio. Japanese immunologist. Studied medicine, PhD in immunology. Positions held at Tokio University and Science University of Tokio. Multiple contributions to suppressor T cells, suppressor T cell factors, I-J expression and restriction (see chapter 11). Philosophical approach to immunology, coorganizer of the Immuno-Semiotics conference with U. Eco (see chapter 12). Also an author of Noh Plays. About Tada: Wada T (2002) A Study on Tomio Tada, *Journal of the Faculty of International Studies of Culture*, Kyushu Sangyo University 21, A1–A11 (in Japanese).

Talmage David W. Studied medicine, MD Washington University St Louis. Positions at University of Colorado and Webb-Waring Lung Institute. Important early contributions to mechanisms of tolerance. Cooperated with F.M. Burnet in the development of the clonal selection theory (see chapter 6). About Talmage: Cruse JM, Lewis RE (1994) David W. Talmage and the advent of the cell selection theory of antibody synthesis. *Journal of Immunology* 153: 919–924.

Tiselius, Arne W. K., 1902–1971. Swedish Chemist. PhD University of Uppsala. Positions at Institute of Advanced Studies, Princeton, USA, and University of Uppsala. Invented electrophoresis, the first procedure by which antibodies were associated with the γ-globulin fraction of serum proteins (see chapter 16). Nobel Prize in Chemistry 1948. About Tiselius: Kekwick RA, Pedersen KO (1974) Arne Tiselius. 1902–1971. *Biographical Memoirs of Fellows of the Royal Society* 20: 401.

Tonegawa Susumu. Born in Japan. PhD University of California San Diego. Postdoctoral training Salk Institute, San Diego. Positions at Basel Institute of Immunology and Massachusetts Institute of Technology, Boston. Discovered immunoglobulin gene rearrangement and their segmental organization in the genome (see chapters 13, 18). Recipient of the 1987 Nobel Prize in Physiology or Medicine for "his discovery of the genetic principle for the generation of antibody diversity".

Uexküll Thure von, 1908–2004. Studied medicine. Residencies at Universities of Hamburg and Charite, Berlin. Positions at University of Munich, University of Giessen. Poineered the field of psychosomatic medicine. Important contributions to biosemiotics (see chapter 12). About Uexküll: Otte R (2001) *Thure von Uexküll. Von der Psychosomatik zur Integrierten Medizin.* Vandenhoeck u. Ruprecht, Göttingen.

Unanue Emil R. Born in Cuba. MD University of Havanna. Postdoctoral training at Scripps Clinic and Research Foundation, La Jolla, University of

Pittsburg, and National Institute of Medical Research, London. Position at Washington University School of Medicine, St. Louis. Important contributions to role of macrophages as antigen-presenting cells and to T cell antigen recognition (see chapter 7).

Uhr Jonathan W. Studied medicine, MD New York University. Residency Mount Sinai Hospital, New York. Postdoctoral training Walter and Eliza Hall Institute, Melbourne. Positions at New York University and Southwestern Medical Center, Dallas, Texas. Important contributions to early studies on antibody responses, immunological memory, 7S inhibition (see chapter 8). Immunological approaches to cancer therapies.

Urbain Jacques. Belgian immunologist. Works at Universite Libre de Bruxelles. Important contributions to research concerning network theory: Experiments to demonstrate production of endogenous anti-idiotypic antibody (Ab2) in mice producing Ab1. Coined the term hypercycle for his observation that injection of an antibody results in the production of endogenous antibodies with similar idiotype (see chapters 12 and 14, interview).

Varela Francisco J., 1946–2001. Born in Chile. Studied biology, University of Chile, Santiago de Chile. PhD in biology at Harvard University, Boston. Professorial positions at University of Chile, University of Colorado, Denver, New York University, Max Planck Institute of Neuroscience, Frankfurt, Centre de Recherche en Epistémologie Appliqué, Paris. Developed, together with H. Maturana, theory of autopoiesis which postulates that living systems, from individual cells to the brain to cognitive societies, are autonomous, self-referential, and self-organizing. Cooperated with A. Coutinho on models of immune networks (see chapters 12, and 14, interview A. Coutinuo). Multiple books, including: Varela FJ (1979) *Principles of Biological Autonomy*. Elsevier North Holland, New York; Varela FJ, Thompson E, Rosch E (1991) *The Embodied Mind: Cognitive Science and Human Experience*. The MIT Press, Cambridge, MA.

Vaz Nelson M. Brazilian immunologist. Research position at University of Belo Horizonte. Multiple publications, including some with A. Coutinho, on self recognition, immune system and individuality, natural autoreactive antibodies, immune networks, evolution of the immune system by exaptation (see chapter 12).

Watson James D. Born in USA. Studied biology at University of Chicago and Indiana University. PhD Indiana University. Postdoctoral training at University of Copenhagen. Positions at Cavendish Laboratory, University of Cambridge, GB, Harvard University, and Cold Spring Harbor Laboratories. Determined the 3-dimensional structure of DNA, together with F. Crick (see chapter 2). Shared with F. Crick and Maurice Wilkins the 1962 Nobel Prize in Physiology or Medicine "for their discoveries concerning the molecular structure of nucleic acids and its significance for information transfer in living material". Autobiography: Watson JD (1980) *The Double Helix: A Personal Account of the Discovery of the Structure of DNA.* Norton Critical Edition, Stent G (ed), Norton, New York.

Wiener Norbert, 1894–1964. Studied mathematics and zoology at Harvard University, philosophy at Cornell University, New York. PhD Harvard University at age 18. Postdoctoral training at Cambridge University, GB, and University of Göttingen, Germany. Positions at Harvard University, General Electric, Massachusetts Institute of Technology. In the context of research on ballistic aiming devices he developed theories on communication and regulation by feedback inhibition which became the basis of cybernetics. Multiple books, including: Wiener N (1948) *Cybernetics: On the Control and Communication in the Animal and the Machine.* MIT Press, Cambridge, MA; Wiener N (1950) *The Human Use of Human Beings.* Da Capo Press; Wiener N (1958) *Nonlinear Problems in Random Theory.* MIT Press & Wiley.

Wigzell Hans. Born in Sweden. Studied medicine, MD Karolinska Institute, Stockholm. Research positions at University of Uppsala and Karolinska Institute. Early contributions to 7S inhibition, discovery of NK cells. Influential experiments on immunoglobulin V region idiotypes and VH linkage of T cell receptors involved in recognition of rat histocompatibility antigens (see chapter 10, and 14, interview). Maternal-fetal idiotypic relationships in immune reactivities, including their role in HIV infection.

Zinkernagel Rolf M. Born in Switzerland. PhD Basel University. Postdoctoral training Australian National University, Canberra. Research positions at Scripps Research Institute, La Jolla, and University of Zurich. Discovered, together with P. Doherty, MHC-restricted recognition of virus-infected cells by cytotoxic T cells (see chapters 7, 10). Multiple contributions to induced tolerance, immunogenicity, anti-viral immunity. Shared with P. Doherty the 1996 Nobel Prize in Physiology or Medicine for "their discoveries concerning the specificity of cell-mediated defense".

Index

active immunity 54, 220
ad hoc hypotheses 11, 221
Adam, G. 93 et seq
adaptive immune system 2, 66, 125 et seq, 225 et seq
afrocentrism 40
agglutinin 54, 215
aldehyde group 218
allotypy 76 et seq, 185, 265
amboceptor 216, 222
amino terminus 55
anti-antibodies 76
antigen-binding site 78
antigenic competition 86
anti-toxin activity 53, 213
arsonate 91, 262
Atkinson, James P. 223
autoimmune diseases 60, 131, 187, 200 et seq
autoimmunity 116, 131, 134, 172
autopoesis 80
auxiliary hypothesis 11, 71
Avery, Oswald Theodore 12 et seq, 55, 166
Avrameas, Stratis 144

B cells 63 et seq, 98, 104 et seq, 126 et seq, 136 et seq, 145 et seq
bacterial toxins 53, 221
bacteriophage 57, 99, 180
Baldamus, W. 27

Baltimore, David 18, 174, 249
Behring, Emil von 53 et seq, 213 et seq, 249
Benacerraf, Baruj 95 et seq, 112, 135, 254
Bence-Jones protein 61
Berlin School of Scientific Philosophy 8
Bertalanffy, Ludwig van 80
Bigelow, Julian 79
black box 15, 99, 108
Bloor, David 11, 250
Bogdanov affair 38 et seq
Bogdanov brothers 38 et seq
Bona, Constantin 134, 250
bone marrow 62 et seq, 74, 145, 182
Bordet, Jules 29, 84, 220, 250
Boyse, Edward "Ted" 108, 250
brain 78 et seq
Bretscher, Peter 68, 70, 87 et seq, 117
Burakoff, Steve 155

Cantor, Harvey 108, 251
carboxy terminus 55
Carnap, Rudolf 8
Cazenave Pierre André 138, 250
Celada, Franco 121
cell activation 59, 102, 126, 142
cell cooperation 63 et seq
central nervous system 79, 119
Chase, Martha 14
Chase, Merill 59, 62
circuit 109 et seq, 112 et seq

Claman, Henry 63, 251
class I antigen 66, 109 et seq, 127 et seq, 149 et seq
class II antigen 66 et seq, 109 et seq, 127 et seq, 149 et seq
clinical research 132, 232
clonal selection theory 1, 19, 58 et seq, 87 et seq, 160 et seq, 165 et seq
Coffman, Robert 126 et seq
cognitive domain 119
Cohn, E.J. 223
Cohn, Melvin 68, 70, 87, 117, 225, 229, 251
colloids 215
complement 29, 108, 218, 220
complement fixation assay 29
Compte, Auguste 8 et seq
connectance 92, 188
constructivism 7, 9 et seq, 39
contrasuppressor circuit 109
Cosenza, Humberto 155
Coutinho Antonio 117, 120, 123, 142, 250
Crick, Francis 14, 18, 55, 251
Croonian lecture 218
crossreactive idiotypic system 90, 98, 165
crystalloid 215
cybernetics 79 et seq, 120, 186, 230
cytokine 71, 126 et seq

Danysz, J. 220
Darwin, Charles 24, 40
Dean, Henry 220
defense system 52, 120
dendritic cells 128, 152
deoxyribonucleic acid (DNA) 12 et seq
detection 10, 30
Deutsch, Ladislav 221
dextran 91
diazobenzaldehyde 218
diazo group 218
differentiation antigen 63
Dobzhansky, Theodosius 237 et seq, 252
Doherty, Peter 67, 100, 252
Dorf, Martin 112
double helix 14, 166
Droege, Wulf 68

Duhem, Pierre Maurice Marie 10 et seq
Duhem-Quine thesis 10, 19

Eco, Umberto 116, 121 et seq, 252
Edelman, Gerald 56, 99, 115, 224, 252
Ehrlich, Paul 29, 54 et seq, 59, 215 et seq, 253
eigen-behavior 85
Eisenberg, Philipp 220
electrophoresis 223, 243
empiricism 8 et seq
19S enhancement 74
epistemology 3 et seq, 10
epitope 84 et seq, 118 et seq
esoteric circle 21 et seq, 73, 225
esoteric community 31
evolution 26, 31, 40, 51 et seq, 237
evolution of species 24, 31, 209
exaptation 118
exoteric community 21
exoteric laity 31

Fab fragments 56
facultative pathogen 51
falsification 18, 24 et seq
Fc fragment 56
feminism 39
Feyerabend, Paul 11, 253
Fleck, Ludwik 4, 26 et seq
formal logic 9, 236, 239
Fujimura, Joan 40, 235, 253

Gauss, Carl Friedrich 40, 235
Germain Ronald 149, 254
Gershon, Richard (Dick) 68, 107 et seq, 112 et seq, 254
Gestalt-vision 23, 103
γ-globulin 55 et seq, 61 et seq, 73 et seq, 97 et seq, 111 et seq, 215 et seq, 223 et seq
Gowans, James L. 59, 254
Graham, Thomas 215
Green, Douglas 232
Greene, Mark 112, 254
Griffith, Frederik 12 et seq, 44

Index

Gross, Paul R. 39 et seq
Guillemin, Roger 35 et seq

haptophore 217 et seq
Haurowitz, Felix 58, 223, 265
Häyry, Pekka 66, 265
healing bodies 214 et seq
heavy and light chains 56, 61, 75, 77, 83, 98
  et seq
Heidelberger, Michael 221, 243, 255
Heisenberg, Werner 239
helper T cells 64, 66 et seq, 97 et seq,
  107 et seq, 128
hematopoietic stem cells 63 et seq
hemolysin 215, 221
Henry, Claudia 74
heredity 12 et seq, 166, 213
Hershey, Alfred 14
Hesse, Mary 11
Hilschmann, Norbert 61, 265
HIV 121, 130, 157, 239
Hoffmann, Erich 29
Hoffmann, G.W. 92, 255
holism 11, 19
Hood, Leroy 112, 256
Hübner, Kurt 11, 256
humanities 8, 38
Hume, David 9 et seq, 234
hybridomas 1, 104, 125, 130

idiotope 76 et seq
idiotypic vaccination 130, 228
idiotypy 74 et seq
I-J antigen 110, 195
I-J region 107, 110
immune bodies 29, 214 et seq
immune response (Ir) genes 66 et seq, 95 et
  seq, 100–103, 111 et seq
immune suppression 106 et seq
immunisin 215
immunoglobulins, see γ-globulin
immunoprecipitation 97
immunotherapy 132, 156, 200
individual antigenic specificity 75 et seq
7S inhibition 73 et seq

innate immune system 128
instructional theory 55 et seq, 228
internal image 78, 85, 119, 129, 230
intravenous immunoglobulin (IVIg)
  therapy 131 et seq
inulin 91

Janeway, Charles A. Jr 113, 128, 256
Jerne, Niels 2, 19, 57 et seq, 74 et seq,
  82 et seq, 226 et seq, 257

Kabat, Elvin 223, 257
Kant, Immanuel 8, 236
Kazatchkine, Michel 131, 137, 257
killer T cell 66 et seq, 107 et seq, 129
Kitasato, Shibasaburo 53, 213, 257
Klein, Jan 113
knockout mice 125 et seq
Knorr Cetina, Karin 34, 211, 258
Koch, Robert 53
Kogon, Eugen 28
Köhler, Georges 1, 104, 226, 258
Köhler, Heinz 155, 258
Koprowski, Hilary 228, 258
Kuhn, Thomas 3, 17 et seq, 207, 258
Kunkel, Henry 61, 76 et seq, 86, 135, 162,
  259

laboratory research 10
laboratory studies 34
Lafitte, Pierre 8
Lakatos, Imre 19, 259
Landsteiner, Karl 55, 59, 86, 221 et seq,
  259
Latour, Bruno 34 et seq, 259
Le Dantec, Félix 220
Lederberg, Joshua 58 et seq, 67, 70, 260
Leibnitz, Gottfried Wilhelm 239
Levitt, Norman 39
Levy, Ron 130, 260
Littré, Émile 8
low zone tolerance 86, 92

Mach, Ernst 8
macrophage 58, 64, 74, 126 et seq

major histocompatibility complex (MHC)
65 et seq, 95 et seq, 127 et seq
mannerism 116 et seq
Marquardt, Martha 217
Matthei, Heinrich 14
Maturana, Humberto 80, 120
Matzinger, Polly 128, 261
McCarthy, Maclyn 12
McDevitt, Hugh 95, 103, 261
Medawar, Peter 60, 63, 207, 261
Mérieux Foundation 228
Meselson, Matthew 14
Metchnikow, Ilya I. 64, 221
Miller, Jacques 63, 261
Mills, John Stuart 8
Milstein, Cesar 1, 19, 104, 226, 261
mimotopes 130
Mitchell, Graham 63
Mitchison, Avrion 59, 62, 122, 262
mixed lymphocyte reaction 65 et seq, 101
Möller, Göran 114, 262
monoclonal antibodies 1, 19, 104 et seq,
125, 151, 229
Morgenroth, Julius 222
Mossman, Tim 126
multicellular organisms 51 et seq
multiple myeloma 61, 76, 131
multispecies system 92, 230
mutation and selection 52
mysticism 78, 230

natural antibody 140
natural objects 10, 36 et seq
natural sciences 8, 12, 38
natural selection 26, 57, 76, 160, 162, 226
negative feedback 76, 80, 120
negative selection 126, 150
neo-positivism 8
neuron 78
Newton, Isaac 8, 11, 20, 22, 234 et seq
Niedermaier, Max 41
Nirenberg, Marshall 14
nitroiodophenyl 91, 99
non-Hodgkin B cell lymphoma 130
Nossal, Gustav 59, 262

nucleotides 14, 236

observation 9, 10
Occam's razor 240
opsonization 64
Oudin, Jacques 74 et seq, 86, 88, 263

paradigma 15, 119
paradigm 21 et seq
paradigm shift 23
paraprotein 75
parasitic microbes 52
paratope 84 et seq
passive immunity 54
Pasteur, Louis 37, 53, 86
Paul, William E. (Bill) 158, 263
Pauling, Linus 58, 223 et seq, 263
peer review 38, 40 et seq, 113
Pernis, Benvenuto 158, 227
phage neutralization 57
phosphorylcholine 90 et seq, 130
plasma cells 59 et seq, 75
pneumococcus 12, 221
Poincaré, Henri 8
polypeptide chains 19, 55, 129
polysaccharide 12, 13, 221
Popper, Karl 3, 17 et seq, 263
Porter, Rodney R. 56, 224, 226, 263
positivism 7 et seq
post-modernism 116
precipitin 54, 215, 220, 222
pre-idea 32, 72
protection bodies 215
proto-idea 72 et seq

quantum logic 239
quantum mechanics 18, 239, 241
Quine, Willar van Orman 10, 15, 264

Raff, Martin 63, 264
Rajewsky, Klaus 123, 162, 264
realism 7 et seq
receptor repertoire selection 126
reductionism 10
regeneration 219

Index

regulatory T cell 114 et seq
Reichenbach, Hans 8, 26, 266
replication 14, 18, 213
reverse transcriptase 18
Richter, Peter H. 92
Rodkey, Scott 84, 266
Rosenblueth, Arturo 80

Sachs, David 167, 265
Schally, Andrew F. 36
Schaudinn, Fritz 29
Schlick, Moritz 8
Schnelle, T. 27
science philosophy 3
science studies 3, 39 et seq
science theory 3, 27, 209
second-order cybernetics 80
self-nonself discrimination 68, 70, 118, 123
self-reference 80, 119
semiosis 117, 122
Sercarz, Eli 121, 123, 170, 265
serum therapy 215 et seq
side chain theory 54 et seq, 218 et seq
Simonsen, Morten 65, 265
Smith, Theobald 53
Snow, C.P. 39
Sokal affair 40
Sokal, Alan 40, 265
Spencer, H. 8
stable state 118
Stahl, Franklin 14
staphylococcal nuclease 91
steric hindrance 77
streptococcal carbohydrate 91
substance of heredity 13, 55
substance sensibilatrice 216
suppressor circuit 109
suppressor factor 107, 110 et seq
suppressor T cell 68, 85, 106 et seq
synapse 78
syphilis 28 et seq

T cell-B cell cooperation 64, 79, 122
T cell receptor idiotypes 98 et seq, 129
T cell receptor repertoire 100, 117

Tada, Tomio 122, 266
Talmage, David 58, 60, 160, 266
TdT 144
Temin, Howard 18
Th1/Th2 paradigm 127
The Central Dogma 18
theory choice 11, 211, 225
thought constraint 31, 216
thought-collective 28 et seq
thought-style 28 et seq
thyrotropin releasing factor (TRF) 35 et
    seq
Tiselius, Arne 223, 266
Toll-like receptor 119, 128
toxoid 218
toxophore 217
transcription 18
transcription factor 115, 119, 138
transformation 13
transforming principle 13
transforming substance 13
translation 18
Treg 114 et seq, 149 et seq
triplet 14, 236

uncertainty relationship 239
underdetermination of theories 11 et seq,
    211
unitarian hypothesis 216
Urbain, Jacques 172, 267

vaccination 53, 78, 129 et seq
vaccines 28, 37, 129 et seq
Varela, Francesco 80, 120, 145, 267
Vaz, Nelson 118 et seq, 267
Verworn, Max 215
VH genes 98, 100 et seq, 107, 112, 129
Vienna Circle 8, 27
Volk, Richard 220

Wassermann reaction 28 et seq, 72
Wassermann, Adolf von 29 et seq
Watson, James 14, 55, 267
Webb, David 232
Weigert, Carl 219

Weigle, William 60
Weiler, E. 93, 100
Weinberg, Steven 40
Weizsäcker, Carl Friedrich von 239
Wells, Harry Gideon 216
Wiener, Norbert 79 et seq, 268
Wigzell, Hans 167, 175, 268
Woit, Peter 42
Woolgar, Steve 35

Zinkernagel, Rolf 67, 100, 103, 268